JT

A New Science of Life

A $10,000 prize has been offered for any completed experiment that documents the validity, or the invalidity, of Rupert Sheldrake's theory.

The prize has been offered by The Tarrytown Group, a membership society affiliated with The Tarrytown Conference Center in Tarrytown, New York. The Group publishes *The Tarrytown Letter,* a monthly newsletter devoted to exploring emerging concepts that suggest large societal change.

Further information may be obtained by writing:

The Sheldrake Prize
Box 222
Tarrytown, New York 10591

The contest closes December 31, 1985.

A New Science of Life

The Hypothesis of Formative Causation

Rupert Sheldrake

J. P. TARCHER, INC.
Los Angeles
Distributed by Houghton Mifflin Company
Boston

Library of Congress Cataloging in Publication Data

Sheldrake, Rupert.
A new science of life.

Bibliography: p. 209
Includes indexes.
1. Biology–Philosophy. I. Title.
QH331.S437 574'.01 81-23270
ISBN 0-87477-221-4 AACR2
ISBN 0-87477-281-8 (ppbk.)

J. P. Tarcher, Inc.
9110 Sunset Blvd.
Los Angeles, CA 90069
Library of Congress Catalog Card No.:

Manufactured in the United States of America

D 10 9 8 7 6 5 4 3 2

First published in Great Britain in 1981 by
Blond & Briggs, Limited

To Dom Bede Griffiths, O.S.B.

Contents

7 The Inheritance of Form

8 The Evolution of Biological Forms

9 Movements and Motor Fields

Preface

MOST BIOLOGISTS take it for granted that living organisms are nothing but complex machines, governed only by the known laws of physics and chemistry. I myself used to share this point of view. But over a period of several years I came to see that such an assumption is difficult to justify. For when so little is actually understood, there is an open possibility that at least some of the phenomena of life depend on laws or factors as yet unrecognized by the physical sciences.

The more I thought about the unsolved problems of biology, the more convinced I became that the conventional approach is unnecessarily restrictive. I started trying to imagine the possible outlines of a broader science of life. In this process the hypothesis put forward in the following pages gradually took shape. Like any new hypothesis, it is essentially speculative, and it will have to be tested experimentally before its value can be judged.

My interest in these problems was stimulated through my association, dating from 1966, with a group of scientists and philosophers engaged in the exploration of areas between science, philosophy and religion. This group, called the Epiphany Philosophers, provided many opportunities for discussions at seminars and informal meetings in Cambridge, and during stays on the Norfolk coast in the Tower Mill at Burnham Overy Staithe. Among the members of this group, I am especially indebted to Professor Richard Braithwaite, Miss Margaret Masterman, the Reverend Geoffrey and Mrs Gladys Keable, Miss Joan Miller, Dr Ted Bastin, Dr Christopher Clarke, and Professor Dorothy Emmet, the editor of *Theoria to Theory*, the group's quarterly journal.

From 1974 to 1978, when I worked in India at the International

Crops Research Institute for the Semi-Arid Tropics, I had many valuable discussions with friends and colleagues in Hyderabad, and the late Mrs J.B.S. Haldane generously gave me the freedom of her large library.

The first draft of this book was written during a stay of a year and a half at Shantivanam Ashram, in the Trichinopoly District of Tamil Nadu. I am grateful to the members of the community for helping to make my stay there so happy, and I owe more than I can say to Dom Bede Griffiths, to whom this work is dedicated. Miss Dina Nanavathy, of the British Council Library in Bombay, kindly kept me supplied with the books I needed.

In the writing and revision of the second draft, after returning to England, I was greatly helped by the advice and encouragement of my friends, and by criticisms and comments from more than fifty people who read the various typescripts. In particular I would like to thank Mr Anthony Appiah, Dr John Beloff, Professor Richard Braithwaite, Dr Keith Campbell, Mrs Jennifer Chambers, Dr Christopher Clarke, the Marchioness of Dufferin and Ava, Professor Dorothy Emmet, Dr Roger Freedman, Dr Alan Gauld, Dr Brian Goodwin, Dr John Green, Mr David Hart, Professor Mary Hesse, Mrs Gladys Keable, Dr Richard LePage, Miss Margaret Masterman, Professor Michael Morgan, Mr Frank O'Meara, Mr Jeremy Prynne, the Honourable Anthony Ramsay, Mrs Jillian Robertson, Dr Tsui Sachs, Professor W.H. Thorpe, F.R.S., Dr Ian Thompson, Mrs R. Tickell (Renée Haynes), Fr E. Ugarte, S.J., and Dr Norman Williams.

I am very grateful to Dr Keith Roberts for doing the drawings and diagrams in this book. Dr Peter Lawrence kindly provided the fruit flies on which the drawings in Figure 17 are based, and Mr Brian Snoad the pea leaves shown in Figure 18.

I thank Mr Mohammed Ibrahim, Mrs Pat Thoburn and Mrs Eithne Thompson for typing the drafts, and Mr Philip Kestelman and Mrs Jenny Reed for their help in reading the proofs.

Hyderabad
March 1981

Introduction

AT PRESENT, the orthodox approach to biology is given by the mechanistic theory of life: living organisms are regarded as physico-chemical machines, and all the phenomena of life are considered to be explicable in principle in terms of physics and chemistry.[1] This mechanistic paradigm[2] is by no means new; it has in fact been predominant for over a century. The main reason why most biologists continue to adhere to it is that it works: it provides a framework of thought within which questions about the physico-chemical mechanisms of life-processes can be asked and answered.

The fact that this approach has resulted in spectacular successes such as the 'cracking of the genetic code' is a strong argument in its favour. Nevertheless, critics have put forward what seem to be good reasons for doubting that all the phenomena of life, including human behaviour, can ever be explained entirely mechanistically.[3] But even if the mechanistic approach were admitted to be severely limited not only in practice but in principle, it could not simply be abandoned; at present it is the only approach available to experimental biology, and will undoubtedly continue to be followed until there is some positive alternative.

Any new theory capable of extending or going beyond the mechanistic theory will have to do more than assert that life involves qualities or factors at present unrecognized by the physical sciences: it will have to say what sorts of things these qualities or factors are, how they work, and what relationship they have to known physico-chemical processes.

The simplest way in which the mechanistic theory could be modified would be to suppose that the phenomena of life depend on a new type of causal factor, unknown to the physical sciences, which interacts with physico-chemical processes within living organisms. Several versions of this vitalist theory have been proposed during the present century,[4] but none has succeeded in making predictions that can be tested, or in suggesting new kinds of experiments. If, to quote Sir Karl Popper, 'the criterion of the scientific status of a theory is its falsifiability, or refutability, or testability',[5] vitalism has so far failed to qualify.

The organismic or holistic philosophy provides a context for what could be a yet more radical revision of the mechanistic theory. This philosophy denies that everything in the universe can be explained from the bottom up, as it were, in terms of the properties of atoms, or indeed of any hypothetical ultimate particles of matter. Rather, it recognizes the existence of hierarchically organized systems which, at each level of complexity, possess properties which cannot be fully understood in terms of the properties exhibited by their parts in isolation from each other; at each level the whole is more than the sum of its parts. These wholes can be thought of as *organisms*, using this term in a deliberately wide sense to include not only animals and plants, organs, tissues and cells, but also crystals, molecules, atoms and sub-atomic particles. In effect this philosophy proposes a change from the paradigm of the machine to the paradigm of the organism in the biological *and* in the physical sciences. In A.N. Whitehead's well-known phrase: 'Biology is the study of the larger organisms, whereas physics is the study of the smaller organisms.'[6]

Various versions of this organismic philosophy have been advocated by many writers, including biologists, for over 50 years.[7] But if organicism is to have more than a superficial influence on the natural sciences, it must be able to give rise to testable predictions. It has not yet done so.[8]

The reasons for this failure are illustrated most clearly in the areas of biology where the organismic philosophy has been most influential, namely embryology and developmental biology. The most important organismic concept put forward so far is that of *morphogenetic fields*.[9] These fields are supposed to help account for, or describe, the coming-into-being of the characteristic forms of embryos and other developing systems. The trouble is that this concept is used ambiguously. The term itself seems to imply the existence of a new type of physical field which plays a role in the development of form. But some organismic theoreticians deny that they are suggesting the existence of any new type of field, entity or factor at present unrecognized by physics;[10] rather, they use this organismic terminology to provide a new way of *talking about* complex physico-chemical systems.[11] This approach seems unlikely to lead very far. The concept of morphogenetic fields can be of practical scientific value only if it leads to testable predictions which differ from those of the conventional mechanistic theory. And such predictions cannot be made unless morphogenetic fields are considered to have measurable effects.

The hypothesis put forward in this book is based on the idea that morphogenetic fields do indeed have measurable physical effects. It proposes that specific morphogenetic fields are responsible for the characteristic form and organization of systems at all levels of complexity, not only in the realm of biology, but also in the realms of chemistry and physics. These fields order the systems with which they are associated by affecting events which, from an energetic point of view, appear to be indeterminate or probabilistic; they impose patterned restrictions on the energetically possible outcomes of physical processes.

If morphogenetic fields are responsible for the organization and form of material systems, they must themselves have characteristic structures. So where do these field-structures come from? The answer suggested is that they are derived from the morphogenetic fields associated with previous similar systems: the morphogenetic fields of all past systems become *present* to any subsequent similar system; the structures of past systems affect subsequent similar systems by a cumulative influence which acts across both space *and time*.

According to this hypothesis, systems are organized in the way they are because similar systems were organized that way in the past. For example, the molecules of a complex organic chemical crystallize in a characteristic pattern because the same substance crystallized that way before; a plant takes up the form characteristic of its species because past members of the species took up that form; and an animal acts instinctively in a particular manner because similar animals behaved like that previously.

The hypothesis is concerned with the *repetition* of forms and patterns of organization; the question of the *origin* of these forms and patterns lies outside its scope. This question can be answered in several different ways, but all of them seem to be equally compatible with the suggested means of repetition.[12]

A number of testable predictions can be deduced from this hypothesis which differ strikingly from those of the conventional mechanistic theory. A single example will suffice: if an animal, say a rat, learns to carry out a new pattern of behaviour, there will be a tendency for any subsequent similar rat (of the same breed, reared under similar conditions, etc.) to learn more quickly to carry out the same pattern of behaviour. The larger the number of rats that learn to perform the task, the easier should it be for any subsequent similar rat to learn it. Thus, for instance, if thousands of rats were trained to perform a new task in a laboratory in London, similar rats should learn to carry out the same task more quickly in laboratories everywhere else. If the speed of learning of rats in another laboratory, say in New York, were to be measured before and after the rats in London were trained, the rats tested on the second occasion should learn more quickly than those tested on the first. The effect should take place in the absence of any known type of physical connection or communication between the two laboratories.

Such a prediction may seem so improbable as to be absurd. Yet, remarkably enough, there is already evidence from laboratory studies of rats that the predicted effect actually occurs.[13]

This hypothesis, called the hypothesis of formative causation, leads to an interpretation of many physical and biological phenomena which is radically different from that of existing theories, and enables a number of well-known problems to be seen in a new light. In the present book, it is sketched out in a preliminary form, some of its implications are discussed, and various ways in which it could be tested are suggested.

Notes

1 For a particularly lucid exposition, see Monod (1972).

2 In the sense of Kuhn (1962).

3 E.g Russell (1945); Elsasser (1958); Polanyi (1958); Beloff (1962); Koestler (1967); Lenartowicz (1975); Popper and Eccles (1977); Thorpe (1978).

4 E.g. Driesch (1908); Bergson (1911a, b). For a discussion of the vitalist approach, see Sheldrake (1980b).
5 Popper (1965), p.37.
6 Whitehead (1928).
7 E.g. Woodger (1929); von Bertalanffy (1933); Whyte (1949); Elsasser (1966); Koestler (1967); Leclerc (1972).
8 In a recent conference on 'Problems of Reduction in Biology', the failure of the organismic approach to make any significant difference to the pursuit of biological research was illustrated by the widespread agreement between mechanists and organicists in *practice*. This led one participant to observe that 'the reductionist/antireductionist arguments among biologists may have as little relevance and impact on the direction of biology as similar arguments conducted in the abstract by philosophers'. (Ayala and Dobzhansky (eds), 1972, p.85).
9 A classical account can be found in Weiss (1939).
10 E.g. Elsasser (1966, 1975); von Bertalanffy (1971).
11 See for example the discussion between C.H. Waddington and R. Thom in Waddington (ed.) (1969), p.242.
12 This point is discussed in the final chapter of the present book.
13 This evidence is discussed in Section 11.2 below.

1 The Unsolved Problems of Biology

1.1 The background of success

The goal of mechanistic biological research was stated particularly clearly over a hundred years ago by T.H. Huxley in the following definition:

> 'Zoological physiology is the doctrine of the functions or actions of animals. It regards animal bodies as machines impelled by various forces and performing a certain amount of work which can be expressed in terms of the ordinary forces of nature. The final object of physiology is to deduce the facts of morphology on the one hand, and those of ecology on the other, from the laws of the molecular forces of matter.'[1]

The subsequent developments of physiology, biochemistry, biophysics, genetics and molecular biology are all foreshadowed in these ideas. In many respects these sciences have been brilliantly successful, none more so than molecular biology. The discovery of the structure of DNA, the 'cracking of the genetic code' and the elucidation of the mechanism of protein synthesis seem to be impressive confirmations of the validity of the mechanistic approach.

The most articulate and influential modern advocates of the mechanistic theory are molecular biologists. Their accounts of the theory usually begin with a brief dismissal of the vitalist and organismic theories. These are defined as survivals of 'primitive' beliefs which are bound to retreat further and further as mechanistic biology advances. The accounts then proceed along the following lines:[2]

The chemical nature of the genetic material, DNA, is now known and so is the genetic code by which it codes for the sequence of amino acids in proteins. The mechanism of protein synthesis is understood in considerable detail. The structure of many proteins has now been worked out. All enzymes are proteins, and enzymes catalyse the complex chains and cycles of biochemical reactions which constitute the metabolism of an organism. Metabolism is controlled by biochemical feedback and several mechanisms are known by which the rates of enzymic activity can be regulated. Proteins and nucleic acids aggregate spontaneously to form structures such as viruses and ribosomes. Given the range of properties of proteins, plus the properties of other physico-chemical systems such as lipid membranes, the properties of living cells can, in principle, be fully explained.

The key to the problems of differentiation and development, about which very little is known, is the understanding of the control of protein synthesis. The way in which the synthesis of certain metabolic enzymes and other proteins is controlled is understood in detail in the bacterium *Escherischia coli.* The control of protein synthesis takes place by more complicated mechanisms in higher organisms, but these should soon be elucidated. Thus differentiation and development should be explicable in terms of series of chemically operated 'switches', which 'switch on' or 'switch off' genes or groups of genes.

The way in which the parts of living organisms are adapted to the functions of the whole, and the apparent purposiveness of the structure and behaviour of living organisms, can be explained in terms of random genetic mutations followed by natural selection, such that those genes which increase the ability of an organism to survive and reproduce will be selected for; harmful mutations will be eliminated. Thus the neo-Darwinian theory of evolution can account for purposiveness; it is totally unnecessary to suppose that any mysterious 'vital factors' are involved.

Very little is known about the functioning of the central nervous system, but eventually the advances of biochemistry, biophysics and electro-physiology should be able to explain what we speak of as the mind in terms of physico-chemical mechanisms in the brain. Thus living organisms are, in principle, fully explicable in terms of physics and chemistry. Our present ignorance about the mechanisms of development and about the central nervous system is due to the enormous complexity of the problems; but now, armed with the powerful new concepts of molecular biology and with the aid of computer models, these subjects can be tackled on a scale and in a way not previously possible.

In the light of past successes, this optimism that *all* the problems of biology can ultimately be solved mechanistically is understandable. But a realistic opinion about the prospects for mechanistic explanation must depend on more than historical extrapolation; it can only be formed after a consideration of the outstanding problems of biology, and of the ways in which they might conceivably be solved.

1.2 The problems of morphogenesis

Biological morphogenesis can be defined as the 'coming-into-being of characteristic and specific form in living organisms.'[3] The first problem is precisely that form comes into being. Biological development is *epigenetic*: new structures appear which cannot be explained in terms of the unfolding or growth of structures which are already present in the egg at the beginning of development.

The second problem is that many developing systems are able to *regulate*; in other words if a part of a developing system is removed (or if an additional part is added), the system continues to develop in such a way that a more or less normal structure is produced. The classical demonstration of this phenomenon was provided in the 1890s in H. Driesch's experiments on sea-urchin embryos. When one of the cells of a very young embryo at the two-celled stage was killed, the remaining cell gave rise not to half a sea-urchin, but to a small but complete sea-urchin. Similarly, small but complete organisms developed after the destruction of any one, two or three cells of embryos at the four-celled stage. Conversely, the fusion of two young sea-urchin embryos resulted in the development of one giant sea-urchin.[4]

Regulation has been demonstrated in many developing systems. However, as development proceeds this capacity is often lost as the fate of different regions becomes determined. But even in systems where determination occurs at an early stage, for example insect embryos, regulation can occur after damage to the egg (Fig. 1).

Results of this type show that developing systems proceed

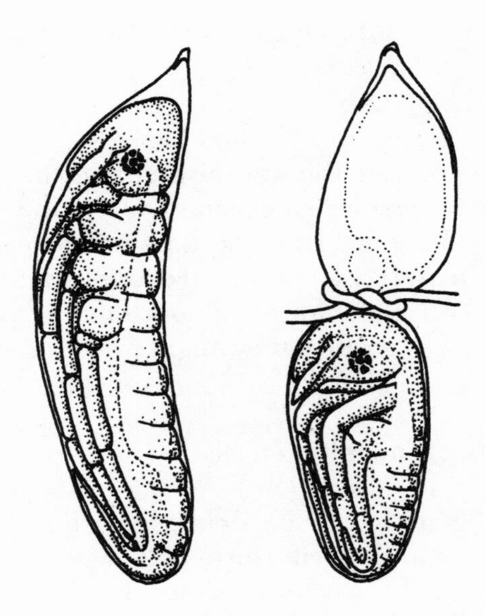

Figure 1 An example of regulation. On the left is a normal embryo of the dragonfly *Platycnemis pennipes*. On the right is a small but complete embryo formed from the posterior half of an egg ligated around the middle soon after laying. (After Weiss, 1939).

towards a morphological goal, and that they have some property which specifies this goal and enables them to reach it even if parts of the system are removed and the normal course of development is disturbed.

The third problem is that of regeneration, whereby organisms are able to replace or restore damaged structures. Plants show an amazing range of regenerative abilities, and so do many of the lower animals; if a flatworm, for example, is cut up into several pieces, each can regenerate into a complete worm. Even many vertebrates possess striking powers of regeneration; for example, if the lens is surgically removed from a newt's eye, a new lens regenerates from the edge of the iris (Fig. 2); in normal embryonic development the lens is formed in a very different way, from the skin. This type of regeneration was first discovered by G. Wolff, who deliberately chose a kind of mutilation which could not have occurred accidentally in nature; there could therefore have been no natural selection for this particular regenerative process.[5]

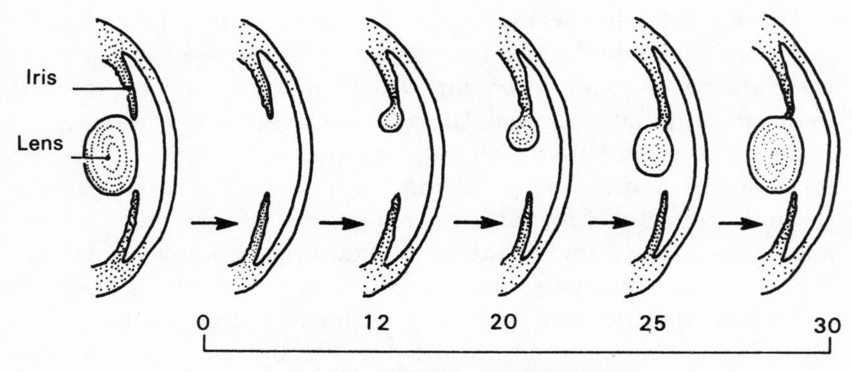

Figure 2 Regeneration of a lens from the margin of the iris in a newt's eye after the surgical removal of the original lens. (Cf. Needham, 1942).

The fourth problem is posed by the simple fact of reproduction: a detached part of the parent becomes a new organism; a part becomes a whole.

The only way in which these phenomena can be understood is in terms of causal entities which are somehow more than the sum of the parts of the developing systems, and which determine the goals of the processes of development.

Vitalists ascribe these properties to *vital factors*, organicists to *morphogenetic fields*, and mechanists to *genetic programmes*.

The concept of genetic programmes is based on an analogy with the programmes that direct the activities of computers. It implies that the fertilized egg contains a pre-formed programme which somehow specifies the organism's morphogenetic goals and co-ordinates and controls its development towards them. But the genetic programme must involve something more than the chemical structure of DNA, because identical copies of DNA are passed on to all cells; if all cells were programmed identically, they could not develop differently. So what exactly is it? In response to this question, the idea can only disintegrate into vague suggestions about physico-chemical interactions somehow structured in time and space; the problem is merely re-stated.[6]

There is a further serious difficulty. A computer programme is put into a computer by an intelligent conscious being, the computer programmer. It is designed and written in order to achieve some definite computational goal. In so far as the genetic programme is regarded as analogous to a computer programme, it implies the existence of some purposive entity which plays the role of the programmer. But if it is argued that genetic programmes are not analogous to ordinary computer programmes, but to those of self-reproducing, self-organising computers, the problem is that such computers do not exist. And even if they did, they would have to have been programmed in the most elaborate way by their inventors to start with. The only way out of this dilemma is to say that the genetic programme has been built up in the course of evolution by a combination of chance mutations and natural selection. But then the similarity to a computer programme disappears and the analogy becomes meaningless.

Orthodox mechanists reject the idea that the seemingly goal-directed behaviour of developing, regulating and regenerating organisms indicates that they are under the control of a vital factor which guides them to their morphological goals. But in so far as mechanistic explanations depend on teleological concepts such as genetic programmes or genetic instructions, goal-directedness can be explained only because it has already been smuggled in. Indeed the properties attributed to genetic programmes are remarkably similar to those with which vitalists endowed their hypothetical vital factors; ironically, the genetic programme seems to be very like a vital factor in a mechanistic guise.[7]

Of course, the fact that biological morphogenesis cannot be explained in a rigorously mechanistic manner at present does not prove that it never will be. The prospects for arriving at such an explanation in the future are considered in the next chapter. But for the time being no conclusive answer is possible.

1.3 Behaviour

If the problems of morphogenesis are dauntingly difficult, those of behaviour are even more so. First, instinct. Consider, for example, how spiders are able to spin webs without learning from other spiders.[8] Or consider the behaviour of European cuckoos. The young are hatched and reared by birds of other species, and never see their parents. Towards the end of the summer, the adult cuckoos migrate to their winter habitat in Southern Africa. About a month later, the young cuckoos congregate together and then they also migrate to the appropriate region of Africa, where they join their elders.[9] They instinctively know that they should migrate and when to migrate; they instinctively recognise other young cuckoos and congregate together; and they instinctively know in which direction they should fly and where their destination is.

Secondly, there is the problem posed by numerous examples of behavioural regulation, in which a more or less normal result is obtained in spite of changes in the parts of the behavioural system. For example, a dog after amputation of a leg regulates its motor activity so that it can walk on three legs. Another dog after the removal of part of a cerebral hemisphere gradually recovers most of its previous abilities. A third dog has obstacles arbitrarily placed in its path. But all three dogs can go from one place to another place they want to get to in spite of disturbances to their motor organs, central nervous system, or environment.

Thirdly, there is the problem of learning and of intelligent behaviour; new patterns of behaviour appear which cannot, apparently, be explained entirely in terms of preceding causes.

An enormous gulf of ignorance lies between all these phenomena and the established facts of molecular biology, biochemistry, genetics and neurophysiology. How, for example, could the migratory behaviour of young cuckoos ultimately be explained in terms of DNA and protein synthesis? Obviously a satisfactory explanation would require more than a demonstration that appropriate genes containing appropriate base-sequences in DNA were *necessary* for this behaviour, or that the behaviour of cuckoos depended on electrical impulses in nerves; it would require some understanding of the connections between specific sequences of bases in DNA, the birds' nervous system, and the migratory behaviour. At present, this connection can only be provided by the same elusive entities that 'explain' all the phenomena of morphogenesis: vital factors, morphogenetic fields or genetic programmes.

Moreover, an understanding of behaviour presupposes an understanding of morphogenesis. For example, even if all the behaviour of a relatively simple lower animal, say a nematode worm, could be understood in detail in terms of the 'wiring' and physiology of its nervous system, there would still be the problem of how the nervous system with this characteristic pattern of 'wiring' came into being in the animal as it developed.

1.4 Evolution

Long before Mendelian genetics was thought of, many distinct varieties and breeds of domesticated animals and plants had been developed by selective breeding. There is no reason to doubt that a comparable development of races and varieties occurs in the wild under the influence of natural rather than artificial selection. The neo-Darwinian theory of evolution claims to be able to explain this type of evolution in terms of random mutations, Mendelian genetics, and natural selection. But even within the mechanistic framework of thought, it is by no means agreed that this type of small-scale or micro-evolution within a species can account for the origin of species themselves, or genera, families and higher taxonomic divisions. One school of thought holds that all large-scale or macro-evolution can be explained in terms of long-continued processes of micro-evolution;[10] the other school denies this, and postulates that major jumps occur suddenly in the course of evolution.[11] But while opinions within mechanistic biology differ as to the relative importance of many small mutations or a few large ones in macro-evolution, there is general agreement that these mutations are random, and that evolution can be explained by a combination of random mutation and natural selection.

However, this theory can never be more than speculative. The evidence for evolution, primarily provided by the Fossil Record, will always be open to a variety of interpretations. For example, opponents of the mechanistic theory can argue that evolutionary innovations are not entirely explicable in terms of chance events, but are due to the activity of a creative principle unrecognized by mechanistic science. Moreover, the selection pressures which arise from the behaviour and properties of living organisms themselves can be considered to depend on an inner organizing factor which is essentially non-mechanistic.

Thus the problem of evolution cannot be solved conclusively.

Vitalist and organismic theories necessarily involve an extrapolation of vitalist and organismic ideas, just as the neo-Darwinian theory involves an extrapolation of mechanistic ideas. This is unavoidable; evolution will always have to be interpreted in terms of ideas which have already been formed on other grounds.

1.5 The Origin of Life

This problem is just as insoluble as that of evolution, for the same reasons. First, what happened in the distant past can never be known for certain; there will probably always be a plethora of speculations on the circumstances of the origin of life on earth. Current ones include the terrestial origin of life within a Primaeval Broth; the infection of the earth by micro-organisms deliberately sent on a space ship by intelligent beings on a planet in another solar system;[12] and the evolution of life on comets containing organic materials derived from interstellar dust.[13]

Secondly, even if the conditions under which life originated could be known, this information would shed no light on the nature of life. Assuming it could be demonstrated, for example, that the first living organisms arose from non-living chemical aggregates, or 'hypercycles' of chemical processes,[14] in a Primaeval Broth, this would not prove that they were entirely mechanistic. Organicists would always be able to argue that new organismic properties emerged, and vitalists that the vital factor entered into the first living system precisely when it first came to life. The same arguments would apply even if living organisms were ever to be synthesized artificially from chemicals in a test tube.

1.6 Limitations of physical explanation

The mechanistic theory postulates that all the phenomena of life, including human behaviour, can in principle be explained in terms of physics. Apart from any problems that might arise from the particular theories of modern physics, or from conflicts between them, this postulate is problematical for at least two fundamental reasons.

First, the mechanistic theory could only be valid if the physical world were causally closed. In relation to human behaviour, this would be the case if mental states either had no reality at all, or were in some sense identical to physical states of the body, or ran parallel to them, or were epiphenomena of them. But if on the other hand the mind were non-physical and yet causally efficacious, capable of *interacting* with the body, then human behaviour could not be fully explained in physical terms. The possibility that mind and body interact is by no means ruled out by the available evidence:[15] at present no clear-cut decision can be made on empirical grounds between the mechanistic theory and the interactionist theory; from a scientific point of view the question remains open. Therefore it is possible that human behaviour, at least, might not be explicable entirely in physical terms, even in principle.

Second, the attempt to account for mental activity in terms of physical science involves a seemingly inevitable circularity, because science itself depends on mental activity.[16] This problem has become apparent within modern physics in connection with the role of the observer in processes of physical measurement; the principles of physics 'cannot even be formulated without referring (though in some versions only implicitly) to the impressions – and thus to the minds – of the observers' (B. D'Espagnat[17]). Thus, since physics presupposes the minds of observers, these minds and their properties cannot be explained in terms of physics.[18]

1.7 Psychology

In psychology, the science of the mind, the problem of the relationship between mind and body can be avoided by ignoring the existence of mental states. This is the approach of the Behaviourist school, which confines its attention only to objectively observable behaviour.[19] But Behaviourism is not a testable scientific hypothesis; it is a methodology. And as an exclusive approach to psychology, its appropriateness is by no means self-evident.[20]

Other schools of psychology have adopted the more straightforward approach of accepting subjective experience as their primary datum. For the purpose of the present discussion, there is no need to consider all the different schools and systems; a single example will suffice to show the biological difficulties raised by a psychological hypothesis developed in an attempt to explain empirical observations. The psychoanalytical schools postulate that many aspects of behaviour and subjective experience depend on the subconscious or unconscious mind. In order to account for the facts of waking experience and of dreams, the unconscious mind has to be endowed with properties totally unlike those of any known mechanical or physical system. In C.G. Jung's development of this concept, it is not even confined to individual minds, but provides a common substratum shared by all human minds, the collective unconscious:

> 'In addition to our immediate consciousness, which is of a thoroughly personal nature and which we believe to be the only empirical psyche (even if we tack on the personal unconscious as an appendix) there exists a second psychic system of a collective, universal, and impersonal nature which is identical in all individuals. This collective unconscious does not develop individually but is inherited. It consists of pre-existent forms, the archetypes, which can only become conscious secondarily and which give definite form to certain psychic contents.'[21]

Jung tried to explain the inheritance of the collective unconscious physically by suggesting that the archetypal forms were 'present in the germplasm'.[22] But it is very doubtful that anything with the properties of the archetypal forms could be inherited chemically in the structure of DNA, or in any other physical or chemical structure in sperm or egg cells. Indeed the idea of the collective unconscious makes little sense in terms of current mechanistic biology, whatever its merits as a psychological theory might be.

However, there is no *a priori* reason why psychological theories should be confined within the framework of the mechanistic theory; they may make better sense in the context of an interactionist theory. Mental phenomena need not necessarily depend on physical laws, but rather follow laws of their own.

The difference between the mechanistic and interactionist approaches can be illustrated by considering the problem of *memory*. According to the mechanistic theory, memories must somehow be stored within the brain. But on an interactionist theory, the mind's properties could be such that past mental states are capable of influencing present states directly, in a manner that does not depend on the storage of physical memory traces.[23] If this were so, a search for physical memory traces within the brain would inevitably be fruitless. And although several different mechanistic hypotheses have been advanced – for example in terms of reverberating circuits of nervous activity, or changes in synaptic connections between nerves, or specific molecules of RNA – there is no persuasive evidence that any of these proposed mechanisms can in fact account for memory.[24]

If memories are not stored physically within the brain, then certain types of memory need not necessarily be confined to individual minds; Jung's notion of an inherited collective unconscious containing archetypal forms could be interpreted as a kind of collective memory.

Such speculations, defensible in the context of interactionism, seem nonsensical from a mechanistic point of view. But the mechanistic theory cannot be taken for granted; at present the idea that all the phenomena of psychology are in principle explicable in terms of physics is itself no more than speculative.

1.8 Parapsychology

In all traditional societies stories are told of men and women with seemingly miraculous powers, and such powers are acknowledged by all religions. In many parts of the world various paranormal abilities are said to be cultivated deliberately within esoteric systems, such as shamanism, sorcery, tantric yoga and spiritualism. And even within modern Western society, there are persistent reports of apparently inexplicable phenomena, such as telepathy, clairvoyance, precognition, memories of past lives, hauntings, poltergeists, psychokinesis, and so on.

Obviously this is an area in which superstition, fraud and credulity are rife. But the possibility that seemingly paranormal events actually occur cannot be dismissed out of hand; the question can only be answered after an examination of the evidence.

The scientific study of allegedly paranormal phenomena has now been going on for almost a century. Although investigators in this field of psychical research have discovered many cases of fraud, and found that some apparently paranormal events can in fact be explained by normal causes, there remains a large body of evidence which seems to defy explanation in terms of any known physical principles.[25] Moreover, numerous experiments designed to test for so-called extra-sensory perception or for psychokinesis have yielded positive results with odds against chance coincidence of thousands, millions or even billions to one.[26]

In so far as these phenomena cannot be explained in terms of the known laws of physics and chemistry, from the conventional mechanistic point of view they ought not to occur.[27] But if they do, then there seem to be two possible types of theoretical approach. The first is to start from the assumption that they depend on laws of physics as yet unknown; the second is to suppose that they depend on non-physical causal factors or connecting principles.[28] Most of the hypotheses of the second type which have been put forward so far have been cast within an interactionist framework. Several recent ones are based on formulations of quantum theory involving 'hidden variables' or 'branching universes', and postulate that mental states play a role in determining the outcomes of probabilistic processes of physical change.[29]

Both the vagueness of such theoretical proposals and the elusiveness of the alleged phenomena make the progress of research in parapsychology very slow. This in turn reinforces the tendency of many mechanistic biologists to ignore or even deny the evidence which seems to show that these phenomena do in fact occur.

1.9 Conclusions

This brief consideration of the outstanding problems of biology does not leave much room for thinking that they can all be solved by an exclusively mechanistic approach. In the case of morphogenesis and animal behaviour the question can be regarded as open; but the problems of evolution and the origin of life are insoluble *per se* and cannot help to decide between the mechanistic and other possible theories of life; the mechanistic theory runs into serious philosophical difficulties in connection with the problem of the limits of physical explanation; in relation to psychology it has no clear advantage over the interactionist theory; and it is in conflict with the apparent evidence for parapsychological phenomena.

On the other hand, although an interactionist approach might be an attractive alternative in the fields of psychology and parapsychology, it has the grave disadvantage of opening up a gulf between psychology and physics. Moreover, its wider biological implications are unclear. For if the interaction of the mind with the body affects human behaviour, then what about the behaviour of other animals? And if a non-physical causal factor plays a part in controlling the behaviour of animals, could it also have a role in the control of morphogenesis? In this case should it be regarded as a factor of the type proposed in vitalistic theories of morphogenesis? If so, in what sense would a vital factor controlling embryological development resemble the human mind?

Thus the interactionist theory, seen in a general biological context, appears to create more theoretical problems than it solves. And it does not seem to lead to any specific testable predictions, apart from allowing for the possibility of parapsychological phenomena.

The organismic approach in its present state also suffers from the disadvantage of suggesting no new lines of empirical research; it offers little more to experimental biology than an ambiguous terminology.

With such feeble alternatives, research in biology will have to continue to follow the mechanistic approach, in spite of its limitations. In this way at least *something* will be found out, even if the major problems of biology remain unsolved. But although in the short term this is the only feasible course of action, looking to the future it seems reasonable to ask whether an alternative is capable of being developed coherently and specifically, and of making testable predictions. If such a theory is to be formulated, the problem of morphogenesis seems to provide the most accessible starting point.

The prospects for improved versions of mechanistic, vitalistic and organismic theories of morphogenesis are discussed in the following chapter.

Notes

1 Huxley (1867), p.74.

2 See for example, Crick (1967) and Monod (1972). Both these authors claim, probably rightly, that their views are representative of those of the majority of their colleagues. In fact, Crick's account, less sophisticated than Monod's, is probably closer to the thinking of most molecular biologists. But Monod's is the clearest and most explicit statement of

the mechanistic position to appear in recent years.
3 Needham (1942), p.686.
4 Driesch (1908).
5 Wolff (1902).
6 Another concept which serves the same explanatory role as the genetic programme is the *genotype*. Though this word is less obviously teleological, it is often used in much the same sense as the genetic programme. In a detailed analysis, Lenartowicz (1975) has shown that if the genotype is simply identified with DNA, its apparent explanatory value disappears.
7 For a fuller discussion, see Sheldrake (1980a).
8 Numerous other examples can be found in von Frisch (1975).
9 Ricard (1969).
10 E.g. Rensch (1959); Mayr (1963); Stebbins (1974).
11 E.g. Goldschmidt (1940); Willis (1940).
12 Crick and Orgel (1973).
13 Hoyle and Wickramasinghe (1978).
14 Eigen and Schuster (1979).
15 See for example the discussions by Beloff (1962) and Popper and Eccles (1977).
16 This problem was pointed out particularly clearly by Schopenhauer (1883).
17 D'Espagnat (1976), p.286.
18 Wigner (1961, 1969).
19 E.g. Watson (1924); Skinner (1938); Broadbent (1961).
20 For critical discussions, see Beloff (1962); Koestler (1967); Popper and Eccles (1977).
21 Jung (1959), p.43.
22 ibid., p.75.
23 Henri Bergson developed an original and stimulating hypothesis of this type in his *Matter and Memory* (1911b). However, other kinds of interactionist hypothesis are possible; for example Beloff (1980) has proposed that the mind interacts with the brain in the *retrieval* of memories, but that the memories themselves are stored as physical traces.
24 A recent review of this subject began as follows: ' "Where or how does the brain store its memories? That is the great mystery." This statement, taken from Boring's (1950) classic work on the history of experimental psychology, is still valid today, despite quarter of a century of intensive work.' (Buchtel and Berlucchi, in Duncan and Weston-Smith (eds), 1977).

But not only is there no evidence that memory traces are stored within the brain, there are also reasons for thinking that no coherent mechanistic explanation of memory in terms of physical traces is possible even in principle (Bursen, 1978).

25 Ashby (1972) provides a critical bibliography covering most aspects of psychical research, and comprehensive reviews of the literature can be found in Wolman (ed.) (1977).

26 For an introductory account, see Thouless (1972).

27 Taylor and Balanovski (1979).

28 For a review of the theoretical literature, see Rao (1977).

29 E.g. Walker (1975); Whiteman (1977); Hasted (1978).

2 Three Theories of Morphogenesis

2.1 Descriptive and experimental research

The description of development can be carried out in many ways: the external form of the developing animal or plant can be drawn, photographed or filmed, providing a series of pictures of its changing morphology; its internal structure, including its microscopic anatomy, can be described at successive stages (Fig. 3); changes in physical quantities such as weight, volume and rate of oxygen consumption can be measured; and changes in the chemical composition of the system as a whole and of regions within it can be analysed. The progressive improvement of techniques permits such descriptions to be made in ever greater detail; for example with the electron microscope the processes of cellular differentiation can be studied at a far higher resolution than with the light microscope, enabling many new structures to be seen; the sensitive analytical methods of modern biochemistry enable changes in concentrations of specific molecules, including proteins and nucleic acids, to be measured in very small samples of tissue; by means of radioactive isotopes, chemical structures can be 'labelled' and 'traced' as the system develops; and techniques for inducing genetic changes in some of the cells of embryos enable their genetically 'marked' descendants to be identified and their fate to be 'mapped'.

The majority of research in embryology and developmental biology is concerned with providing factual descriptions by means of such techniques; these descriptions are then classified and compared in order to establish how different sorts of changes are correlated within a given system, and in what ways different systems resemble each other. These purely descriptive results cannot in themselves lead to an understanding of the causes of development,

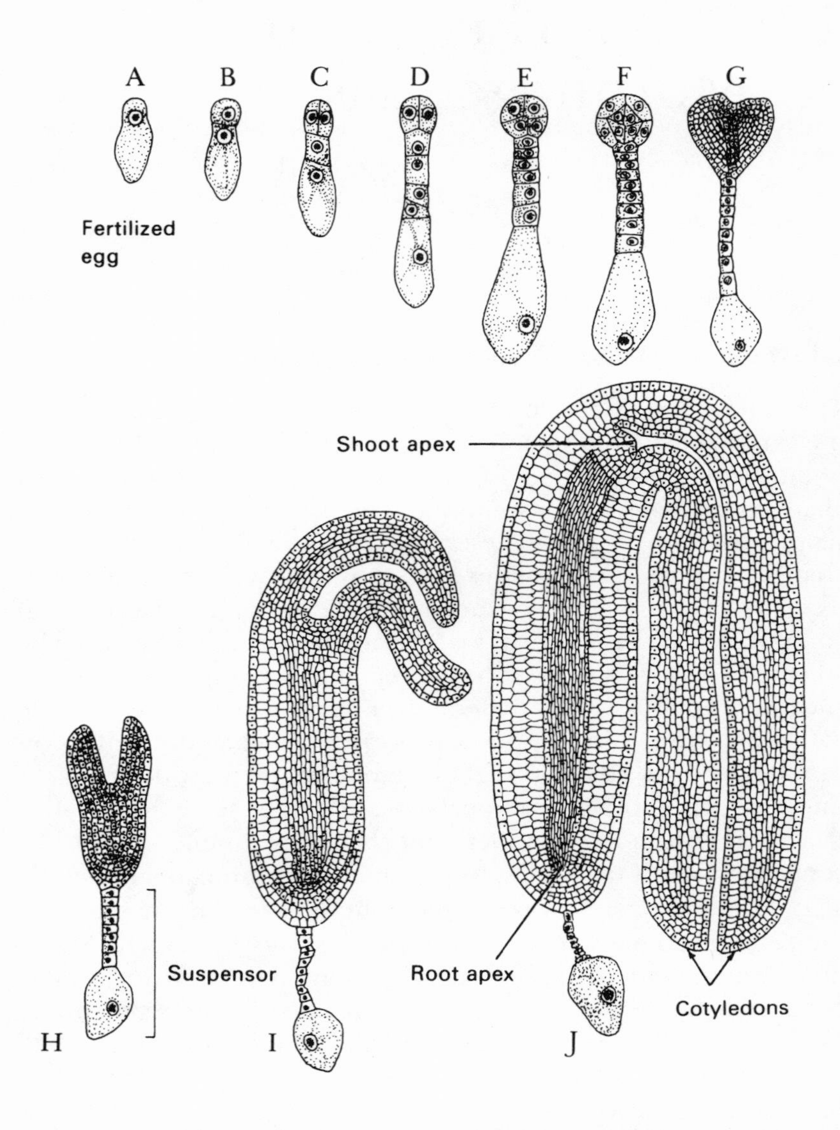

Figure 3 Stages in the development of the embryo of the shepherd's purse plant, *Capsella bursa-pastoris.* (After Maheshwari, 1950).

although they may suggest hypotheses.[1] The latter can then be investigated by means of experimental perturbations of development: for example, the environment can be changed; physical or chemical stimuli can be applied to specific places on or in the system; parts of the system can be removed and their development studied in isolation; the reaction of the system to the removal of parts can be observed; and the effects of combining different parts can be studied by grafts and transplantations.

The major problems thrown up by this type of research have been summarized in Section 1.2: biological development is epigenetic, or in other words involves an increase in complexity of form and organization which cannot be explained in terms of the unfolding or decomposition of a pre-formed but invisible structure; many developing systems can regulate, i.e. produce a more or less normal structure if part of the system is destroyed or removed at a sufficiently early stage; many systems can regenerate or replace missing parts; and in vegetative and sexual reproduction, new organisms are formed from detached parts of parent organisms. One further important generalization is that in developing systems the destiny of cells and tissues is determined by their position within the system.

Mechanistic, vitalist and organismic theories all start from this established body of facts and results, about which there is general agreement, but they differ radically in their interpretations.

2.2 Mechanism

The modern mechanistic theory of morphogenesis ascribes a role of prime importance to DNA, for four main reasons. First, many cases of hereditary differences between animals or plants of a given species have been found to depend on genes, which can actually be 'mapped' and located at particular places on particular chromosomes. Secondly, the chemical basis of genes is known to be DNA and their specificity is known to depend on the sequence of purine and pyrimidine bases in the DNA. Thirdly, it is known how DNA is able to act as the chemical basis of heredity: on the one hand, it serves as a template for its own replication, owing to the specificity of the pairing of the bases in its two complementary strands; on the other hand it serves as the template for the sequence of amino acids in proteins. It does not play the latter role directly; one of its strands is first 'transcribed' to give a single-stranded molecule of 'messenger' RNA from which, in the process of protein synthesis, the sequence of bases is 'read off' three at a time. Different triplets of bases specify different animo acids, and thus the genetic code is 'translated' into a sequence of amino acids, which are linked together to give characteristic polypeptide chains, which then fold up to give proteins. Finally, the characteristics of a cell depend on its proteins: its metabolism and its capacities for chemical synthesis on enzymes, some of its structures on structural proteins, and the surface properties which enable it to be 'recognized' by other cells on special proteins on its surface.

Within the mechanistic framework of thought, the central problem of development and morphogenesis is seen as the control of protein synthesis. In bacteria, specific chemicals called inducers can cause specific regions of the DNA to be transcribed into messenger RNA, on which template specific proteins are then made. The classic example is the induction of the enzyme ß-galactosidase by lactose in *Escherischia coli*. The 'switching on' of the gene takes place through a complicated system involving a repressor protein which blocks transcription by combining with a specific region of the DNA; its tendency to do so is greatly reduced in the presence of the chemical inducer. By a comparable process, specific chemical repressors can 'switch off' genes. In animals and plants the system by which genes are 'switched' on and off is more complicated, and not at present understood. Further complications arise from the recently-discovered fact that messenger RNA can be made up of pieces transcribed from different regions of the DNA, and subsequently joined together in a specific way. Moreover, the synthesis of proteins is also controlled at the 'translational level'; protein synthesis can be 'switched' on and off by a variety of factors even in the presence of appropriate messenger RNA.

The different proteins made by different types of cells thus depend on the way in which protein synthesis is controlled. The only way in which this can be understood mechanistically is in terms of physico-chemical influences on the cells; patterns of differentiation must therefore depend on physico-chemical patterns within the tissue. The nature of these influences is not known, but various possibilities have been suggested: concentration gradients of specific chemicals; 'diffusion-reaction' systems with chemical feedback; electrical gradients; electrical or chemical oscillations; mechanical contacts between cells; or various other factors, or combinations of different factors. The cells must then respond to these differences in characteristic ways. One current way of thinking about this problem is to regard these physical or chemical factors as providing 'positional information' which the cells then 'interpret' in accordance with their genetic programme by 'switching on' the synthesis of particular proteins.[2]

These various aspects of the central problem of the control of protein synthesis are at present under active investigation. Most mechanistic biologists hope that the solution of this problem will provide, or at least lead towards, an explanation of morphogenesis in purely mechanistic terms.

In order to assess whether such a mechanistic explanation of morphogenesis is likely, or even possible, a number of difficulties will have to be considered one by one:

(i) The explanatory role ascribed to DNA and the synthesis of specific proteins is severely restricted in its scope by the fact that both the DNA and the proteins of different species may be very similar. For example, in a detailed comparison of human and chimpanzee proteins, a considerable number have been found to be identical, and others to differ only slightly: 'Amino acid sequencing, immunological and electrophoretic methods yield concordant estimates of genetic resemblance. These approaches all indicate that the average human polypeptide is more than 99 percent identical to its chimpanzee counterpart'.[3] Comparisons of the so-called non-repeated DNA sequences (i.e. those parts believed to be of genetic significance) show that the overall difference between the DNA sequences of humans and chimpanzees is only 1.1 percent.

Similar comparisons between different species of mice or of the fruit fly *Drosophila* have revealed *larger* differences between these closely related species than between humans and chimpanzees, leading to the conclusion that 'the contrasts between organismal and molecular evolution indicate that the two processes are to a large extent independent of each other.'[4]

However, assume for the purpose of argument that the hereditary differences between species as different as humans and chimpanzees can indeed be explained in terms of very small changes in protein structure, or small numbers of different proteins, or genetic changes which affect the control of protein synthesis (perhaps depending to some extent on differences of arrangement of DNA within the chromosomes), or combinations of these factors.

(ii) Within the same organism, different patterns of development take place while the DNA remains the same. Consider, for example, the arm and the leg of a man: both contain identical cell types (muscle cells, connective tissue cells, etc.) with identical proteins and identical DNA. So the differences between the arm and the leg cannot be ascribed to DNA *per se*; they must be ascribed to pattern-determining factors which act differently in the developing arm and leg. The precision of arrangement of the tissues – for example the joining of tendons to the right parts of the bones – shows that these pattern-determining factors must work with great precision. The mechanistic theory of life means that these factors must be regarded as physico-chemical in nature. However, their nature is at present unknown.

(iii) Even if physical or chemical factors determining a pattern of differentiation can be identified, there is still the problem of how these factors are themselves patterned in the first place. This problem can be illustrated by considering two of the very few cases in which chemical 'morphogens' have actually been isolated.

First, in the cellular slime moulds, free-living amoeboid cells aggregate together under certain conditions to form a 'slug' which, after moving around for some time, grows up into the air and differentiates into a stalk bearing a spore-mass (Fig. 4). The aggregation of these cells have been shown to depend on a relatively simple chemical, cyclic AMP (adenosine 3′, 5′ – monophosphate). But in the composite organism, although the distribution of cyclic AMP is related to the pattern of differentiation, 'it is not clear whether the cyclic AMP pattern is a cause or consequence of prestalk-prespore differentiation'. Moreover, even if it does play a causal role in differentiation, it cannot itself account for the pattern in which it is distributed, nor for the fact that this pattern varies

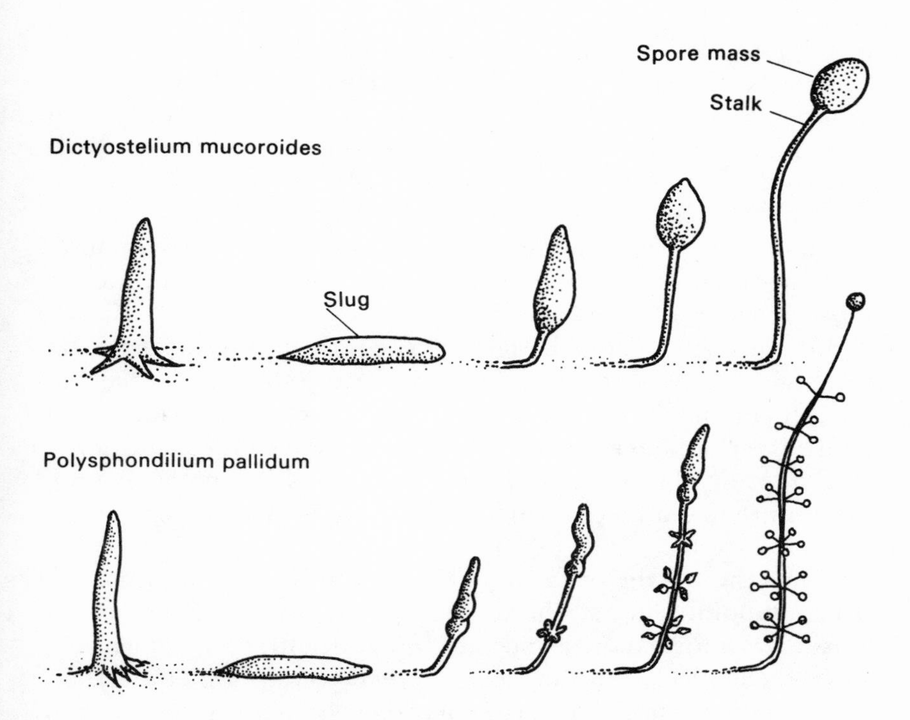

Figure 4 The migration and culmination stages of two species of slime mould. On the left are the newly-developed composite organisms, formed by the aggregation of numerous free-living amoeboid cells. These migrate as 'slugs', and then grow upwards, differentiating into stalks bearing spore bodies. (After Bonner, 1958).

from species to species: some other factors must be responsible for its patterned distribution. There is a wide variety of opinion on the possible nature of these factors.[5]

Secondly, in higher plants the hormone auxin (indolyl – 3 – acetic acid) is known to play a role in the control of vascular differentiation. But then what controls the production and distribution of auxin? The answer seems to be: vascular differentiation itself. Auxin is probably released by differentiating vascular cells as a by-product of the protein breakdown which occurs as the cells mature. Thus the system is circular: it helps to maintain patterns of differentiation, but it does not explain how they are established to start with.[6]

However, assume for the purpose of argument that it might be possible to identify what factors give a pattern to the physical or chemical influences which in turn determine the pattern of differentiation; assume also that the ways in which these controlling factors are themselves controlled can be identified; and so on. Now there is the problem of regulation: if part of the system is removed, this complicated series of physico-chemical patterns must be disrupted. But somehow the remaining parts of the system manage to change their usual course of development and go on to produce a more or less normal end result.

This problem is generally agreed to be extremely difficult; it is far from being understood even in outline. Supporters of the mechanistic theory hope that it can be solved through much painstaking effort; their opponents deny that it can be solved mechanistically even in principle. However, for the purpose of argument, assume once more that a mechanistic solution can be achieved.

(iv) Then there is the problem of how this 'positional information' brings about its effects. The simplest possibility would be that the 'positional information' is specified by a concentration gradient of a specific chemical, and that cells exposed to more than a certain concentration synthesize one set of proteins, while cells exposed to concentrations below this threshold synthesize another set of proteins. Again, assume that this or other mechanisms by which 'positional information' can be 'interpreted' could actually be identified.[7] Now, at the end of this chain of highly optimistic assumptions, the situation is reached in which different cells arranged in a suitable pattern make different proteins.

So far, there has been a set of one-to-one relationships: a gene is 'switched on' by a specific stimulus; the DNA is transcribed into RNA; and the RNA is translated into a particular sequence of amino acids, a polypeptide chain. But now this simple causal sequence comes to an end. How do the polypeptide chains fold up into the characteristic three-dimensional structures of proteins? How do the proteins give the cells their characteristic structures? How do cells aggregate together to give tissues of characteristic structures? And so on. These are the problems of morphogenesis proper: the synthesis of specific polypeptide chains provides the basis for the metabolic machinery and the structural materials on which morphogenesis depends. The polypeptide chains and the proteins into which they fold up are undoubtedly necessary for morphogenesis; but what actually determines the patterns and structures into which the proteins, cells and tissues combine? The mechanistic assumption is that all this can be explained in terms of physical interactions, and that it takes place spontaneously, given the right proteins in the right places at the right times and in the right sequence. At this crucial stage, mechanistic biology effectively abdicates, and the problem of morphogenesis is simply left to physics.

It is indeed the case that polypeptide chains fold up spontaneously, given the right conditions, into proteins of characteristic three-dimensional structure. They can even be made to unfold and then, by changing the conditions, to fold up again in test tubes, so this process does not depend on any mysterious property of living cells. Moreover protein sub-units can aggregate together under test-tube conditions to form structures which are normally produced inside living cells: for example, sub-units of the protein tubulin join together into long rod-like structures called microtubules.[8] And yet more complex structures, such as ribosomes, can be formed by the spontaneous aggregation of various protein and RNA components. Other classes of substances, for example the lipids of cell membranes, can also aggregate together spontaneously in the test tube.

In so far as these structures undergo spontaneous self-assembly, they resemble crystals; many of them can indeed be regarded as crystalline or quasi-crystalline. So in principle they pose no more, or no less, of a problem than normal crystallization; the same sorts of physical process can be assumed to be at work.

Nevertheless, by no means all morphogenetic processes can be regarded as types of crystallization. They must involve a number of other physical factors; for example, the shapes taken up by membranes must be influenced by the forces of surface tension, and the structures of gels and sols by the colloidal properties of their constituents. And then some of the patterns may arise from statistically random fluctuations; simple examples of the appearance of 'order through fluctuations' have begun to be studied from the point of view of irreversible or non-equilibrium thermodynamics in inorganic systems, and comparable processes may well be at work within cells and tissues.[9]

However, the mechanistic theory does not merely suggest that these and other physical processes play a part in morphogenesis; it asserts that morphogenesis is entirely explicable in terms of physics. What does this mean? If everything observable is *defined* as being physically explicable in principle, just because it happens, then it must be so by definition. But this does not necessarily mean it can be explained in terms of the *known* laws of physics. In relation to biological morphogenesis this explanation could be said to have been achieved if a biologist who was supplied with the entire sequence of bases in the DNA of an organism and a detailed description of the physico-chemical state of the fertilized egg, and of the environment in which it developed, could *predict* in terms of the fundamental laws of physics (e.g. quantum field theory, the equations of electro-magnetism, the second law of thermodynamics, etc.) first, the three-dimensional structure of all the proteins the organism would make; secondly, the enzymic and other properties of these proteins; thirdly, the organism's entire metabolism; fourthly, the nature and consequences of all the types of positional information that would arise during its development; fifthly, the structure of its cells, tissues and organs and the form of the organism as a whole; and finally, in the case of an animal, its instinctive behaviour. If all these predictions could be made successfully, and if, in addition, the course of processes of regulation and regeneration could also be predicted *a priori*, this would indeed be a conclusive demonstration that living organisms are fully explicable in terms of the known laws of physics. But, of course, nothing of the sort can be done at present. So there is no way of demonstrating that such an explanation is possible. It might not be.

Thus if the mechanistic theory states that all the phenomena of morphogenesis are capable in principle of being explained in terms of the known laws of physics, it might well be wrong: so little is understood at present that there seem to be no good grounds for a firm belief in the adequacy of the known laws to explain all the phenomena. But at any rate this is a testable theory; it could be refuted by the discovery of a new law of physics. If on the other hand the mechanistic theory states that living organisms obey both known and unknown laws of nature, then it would be irrefutable; it would simply be a general statement of faith in the possibility of explanation. It would not be opposed to organicism and vitalism; it would include them.

In practice, the mechanistic theory of life is not treated as a rigorously defined, refutable scientific theory; rather, it serves to provide a justification for the conservative method of working within the established framework of thought provided by existing physics and chemistry. Although it is usually understood to mean that living organisms are in principle fully explicable in terms of the known laws of physics, if a new law of physics were to be discovered, and thus became known, the mechanistic theory could easily be modified to include it. Whether this modified theory of life were to be called mechanistic or not would only be a matter of definition.

When so little is understood about the phenomena of morphogenesis and behaviour, the possibility can by no means be ruled out that at least some of them depend on a causal factor as yet unrecognized by physics. In the mechanistic approach, this question is simply put aside. Nevertheless it remains entirely open.

2.3 Vitalism

Vitalism asserts that the phenomena of life cannot be fully understood in terms of physical laws derived only from the study of inanimate systems, but that an additional causal factor is at work in living organisms. A typical statement of a nineteenth century vitalist position was made by the chemist Liebig in 1844: he argued that although chemists could already produce all sorts of organic substances, and would in future produce many more, chemistry would never be in a position to create an eye or a leaf; besides the recognized causes of heat, chemical affinity, and the formative force of cohesion and crystallization 'in living bodies there is added yet a fourth cause which dominates the force of cohesion and combines the elements in new forms so that they gain new qualities – forms and qualities which do not appear except in the organism.'[10]

Ideas of this type, although widely held, were too vague to provide an effective alternative to the mechanistic theory. It was only at the beginning of this century that neo-vitalist theories were worked out in some detail. In relation to morphogenesis, the most important was that of the embryologist Hans Driesch. If a modern vitalist theory were to be developed, Driesch's would provide the best foundation on which to build.

Driesch did not deny that many features of living organisms could be understood in physico-chemical terms. He was well aware of the findings of physiology and biochemistry, and of the potential for future discovery: 'There are many specific chemical compounds present in the organism, belonging to the different classes of the chemical system, and partly known in their constitution, partly unknown. But those that are not yet known will probably be known some day in the near future, and certainly there is no theoretical impossibility about discovering the constitution of albumen [protein] and how to "make" it.'[11] He knew that enzymes ('ferments') catalysed biochemical reactions and could do so in test tubes: 'There is no objection to our regarding almost all metabolic processes inside the organism as due to the intervention of ferments or catalytic materials, and the only difference between inorganic and organic ferments is the very complicated character of the latter and the very high degree of their specification.'[12] He knew that Mendelian genes were material entities located in the chromosomes, and that they were probably chemical compounds of specific structure.[13] He thought that many aspects of metabolic regulation and physiological adaptation could be understood along physico-chemical lines[14] and that there were in general 'many processes in the organism . . . which go on teleologically or purposefully on a fixed machine-like basis'.[15] His opinions on these subjects have been confirmed by the subsequent advances of physiology, biochemistry, and molecular biology. Obviously, Driesch was unable to anticipate the details of these discoveries, but he regarded them as possible and in no way incompatible with vitalism.

In relation to morphogenesis, he considered that 'it must be granted that a machine, as we understand the word, might very well be the motive force of organogenesis in general, if only normal, that is to say, if only undisturbed development existed, and if taking away parts of our system led to fragmental development.'[16] But, in fact, in many embryonic systems removal of a part of the embryo is followed by a process of regulation, whereby the remaining tissues reorganize themselves and go on to produce an adult organism of more or less normal form.

The mechanistic theory has to attempt to account for development in terms of complex physical or chemical interactions between the parts of the embryo. Driesch argued that the fact of regulation made any such machine-like system inconceivable, because the system was able to remain a whole and produce a typical final result, whereas no complex three-dimensional machine-like system could remain a whole after the arbitrary removal of parts.

This argument is open to the objection that it is, or will be at some time in the future, invalidated by advances in technology. But at least it does not seem to have been refuted so far. For example, although computerized cybernetic systems can respond appropriately to certain types of functional disturbance, they do so on the basis of a fixed structure. They cannot regenerate their own physical structure; for example, if parts of the computer are destroyed at random, they cannot be replaced by the machine itself, nor can the system go on functioning normally after the arbitrary removal of parts. The other item of modern technology which might seem relevant is the hologram, from which pieces can be removed but which can still give rise to a complete three-dimensional image. However, the hologram can only do so when it is part of a larger functional whole, including a laser, mirrors, etc. These structures cannot be regenerated after arbitrary damage, for instance if the laser is smashed.

Driesch believed that the facts of regulation, regeneration and reproduction showed that there was something about living organisms which remained a whole even though parts of the physical whole could be removed; it acted *on* the physical system but was not itself part of it. He called this non-physical causal factor *entelechy*. He postulated that entelechy organized and controlled physico-chemical processes during morphogenesis; the genes were responsible for providing the material *means* of morphogenesis – the chemical substances to be ordered – but the ordering itself was brought about by entelechy. Clearly morphogenesis could be *affected* by genetic changes which changed the means of morphogenesis, but this would not prove that it could be *explained* simply in terms of genes and the chemicals to which they gave rise. Similarly, the nervous system provided the means for the actions of an animal, but entelechy organised the activity of the brain, using it as an instrument, as a pianist plays on a piano. Again, behaviour could be affected by damage to the brain, just as the music played by the pianist would be affected by damage to the piano; but this would only prove that the brain was a necessary means for behaviour, as the piano is a necessary means for the pianist.

Entelechy is a Greek word whose derivation (en-telos) indicates something which bears its end or goal in itself; it 'contains' the goal towards which a system under its control is directed. Thus if a normal pathway of development is disturbed, the system can reach the same goal in a different way. Driesch considered that development and behaviour were under the control of a hierarchy of entelechies, which were all ultimately derived from, and subordinated to, the overall entelechy of the organism.[17] As in any hierarchical system, such as an army, mistakes were possible and entelechies might behave 'stupidly', as they do in cases of super-regeneration, when a superfluous organ is produced.[18] But such stupidities do not disprove the existence of entelechy any more than military errors disprove that soldiers are intelligent beings.

Driesch described entelechy as an 'intensive manifoldness', a non-spatial causal factor which nevertheless acted into space. He emphasized that it was a natural (as opposed to a metaphysical or mystical) factor which acted on physico-chemical processes. It was not a form of energy, and its action did not contradict the second law of thermodynamics or the law of conservation of energy. Then how did it work?

Driesch was writing during the era of classical physics, when it was generally thought that all physical processes were fully deterministic, in principle completely predictable in terms of energy, momentum, etc. But he considered that physical processes could not be fully determinate, because otherwise the non-energetic entelechy could not act upon them. He therefore concluded that, at least in living organisms, microphysical processes were not fully determined by physical causality, although, on average, physico-chemical changes obeyed statistical laws. He suggested that entelechy acted by affecting the detailed *timing* of microphysical processes, by 'suspending' them and releasing them from suspension whenever required for its purposes:

> 'This faculty of a temporary suspension of inorganic becoming is to be regarded as the most essential ontological characteristic of entelechy . . . Entelechy, according to our view, is quite unable to remove any kind of "obstacle" to happening . . . for such a removal would require energy, and entelechy is non-energetic. We only admit that entelechy may set free into

actuality what it has *itself* prevented from actuality, what it has suspended hitherto.'[19]

Although this bold proposal of a physical indeterminism within living organisms seemed to be completely unacceptable from the point of view of deterministic classical physics, it seems much less outrageous in the light of quantum theory. Heisenberg deduced the uncertainty principle in 1927, and it soon became clear that positions and timings of microphysical events could be predicted only in terms of probabilities. By 1928, the physicist Sir Arthur Eddington was able to speculate that the mind influences the body by affecting the configuration of quantum events within the brain through a causal influence on the probability of their occurrence. 'Unless it belies its name, probability can be modified in ways which ordinary physical entities would not admit of.'[20] Comparable ideas have been proposed by the neurophysiologist Sir John Eccles, who summarised his suggestion as follows:

'The neurophysiological hypothesis is that the "will" modifies the spatio-temporal activity of the neuronal network by exerting spatio-temporal "fields of influence" that become effective through this unique detector function of the active cerebral cortex. It will be noted that the "will" or "mind influence" has itself some spatio-temporal patterned character in order to allow it this operative effectiveness.'[21]

More recently a number of similar but more detailed proposals have been put forward by physicists and by parapsychologists[22] (Section 1.8).

In line with these proposals, a modern vitalist theory could be based on the hypothesis that entelechy, to use Driesch's terminology, orders physico-chemical systems by influencing physically indeterminate events within the statistical limits set by energetic causation. To do so, it must itself be patterned spatio-temporally.

But then how does entelechy come to have this patterned character? A possible answer is suggested by the interactionist theory of memory outlined in Section 1.7. If memories are not stored physically within the brain, but somehow involve a direct action across time,[23] then they need not be confined to individual brains; they could pass from person to person, or a sort of 'pooled' memory could be inherited from countless individuals in the past.

These ideas can be generalized to include the instincts of animals.

Instincts could be inherited through the collective memory of the species; an instinct would be like a habit acquired not just by individuals but by the species as a whole.

Ideas of this type have already been proposed by a number of authors;[24] for example, the psychical researcher W. Carington suggested that instinctive behaviour such as the web-spinning of a spider 'may be due to the individual creature (e.g. spider) being linked up into a larger system (or common subconscious if your prefer it) in which all the web-spinning experience of the species is stored up.'[25] The zoologist Sir Alister Hardy developed this idea by suggesting that this shared experience would act as a sort of 'psychic blueprint':

> 'There would be two parallel streams of information – the DNA code supplying the varying physical form of the organic stream to be acted upon by selection – and the psychic stream of shared experience – the subconscious species "blueprint" –which together with the environment, would select those members in the population better able to carry on the race.'[26]

In these suggestions, the type of inheritance depending on a non-physical memory-like process is confined to the realm of behaviour. A further generalization of this idea to include the inheritance of form would bring it into contact with Driesch's concept of entelechy: the characteristic pattern imposed on a physico-chemical system by entelechy would depend on the spatio-temporal patterning of entelechy itself by a sort of memory-process. A sea urchin embryo, for example, would develop as it did because its entelechy contained the 'memory' of the developmental processes of all previous sea urchins; moreover the 'memory' of the larval and adult forms of previous sea urchins would enable entelechy to direct development towards these normal goals even if the embryo was damaged, accounting for regulation.

Thus a possible vitalist theory of morphogenesis could be summarised as follows: the genetic inheritance of DNA specifies all the possible proteins the organism can make. But the organisation of the cells, tissues and organs, and the co-ordination of the development of the organism as a whole, is determined by entelechy. The latter is inherited non-materially from past members of the same species; it is not a type of matter or energy, although it acts upon the physico-chemical systems of the organism under its control. This action is possible because entelechy acts as a set of 'hidden variables' which influence probabilistic processes.

This theory is by no means vacuous, and could probably be tested experimentally; but it seems fundamentally unsatisfactory simply because it is vitalistic. Entelechy is essentially non-physical, by definition; even though it could, *ex hypothesi*, act on material systems by providing a set of variables which from the point of view of quantum theory are hidden, this would still be an action of unlike on unlike. The physical world and the non-physical entelechy could never be explained or understood in terms of each other.

This dualism, inherent in all vitalist theories, seems particularly arbitrary in the light of the discoveries of molecular biology of the 'self-assembly' of structures as complex as ribosomes and viruses, indicating a difference of degree, and not of kind, from crystallization. Although the self-organisation of living organisms as a whole is more complex than that of ribosomes or viruses, and produces a far greater internal heterogeneity, there is sufficient similarity to suggest that here again is a difference of degree. This, at any rate, is what both mechanists and organicists prefer to think.

Possibly a vitalist theory would have to be accepted if no other satisfactory explanation of the phenomena of life were conceivable. In the early part of this century when vitalism seemed to be the only alternative to the mechanistic theory, it gained considerable support in spite of its essential dualism. But the development of the organismic theory over the last 50 years has provided another possibility which, by incorporating many aspects of vitalism within a larger perspective, has effectively superseded it.

2.4 Organicism

Organismic theories of morphogenesis have developed under a variety of influences: some from philosophical systems, especially those of A.N. Whitehead and J.C. Smuts; some from modern physics, in particular from the field concept; others from Gestalt psychology, itself strongly influenced by the concept of physical fields; and some from the vitalism of Driesch.[27] These theories deal with the same problems that Driesch claimed were insoluble in mechanistic terms – regulation, regeneration and reproduction – but whereas Driesch proposed the non-physical entelechy to account for the properties of wholeness and directiveness exhibited by developing organisms, organicists proposed morphogenetic (or embryonic, or developmental) *fields*.

This idea was put forward independently by A. Gurwitsch in 1922[28] and P. Weiss in 1926.[29] However, apart from stating that morphogenetic fields played an important role in the control of morphogenesis, neither of these authors specified what they were or how they worked. The field terminology was soon taken up by other developmental biologists, but it remained ill-defined, although it served to suggest analogies between properties of living organisms and inorganic electro-magnetic systems. For example, if an iron magnet is cut into two parts, two whole magnets are produced, owing to the properties of the magnetic field; similarly, the morphogenetic field was supposed to account for the 'wholeness' of detached parts of organisms which were capable of growing into new organisms.

C.H. Waddington suggested an extension of the idea of the morphogenetic field to take into account the temporal aspect of development. He called this new concept the *chreode* (from the Greek chrē, it is necessary, and hodos, route or path) and illustrated it by means of a simple three-dimensional 'epigenetic landscape' (Fig. 5).[30] In this model the path followed by the ball as it rolls downwards corresponds to the developmental history of a particular part of an egg. As embryology proceeds there is a branching series of alternative paths represented by the valleys. These correspond to the pathways of development of the different types of organ, tissue and cell. In the organism these are quite distinct; for example, the kidney and liver have definite structures and do not grade into each other through a series of intermediate forms. Development is *canalized* towards definite end points. Genetic changes or environmental perturbations may push the course of development (represented by the pathway followed by the ball) away from the valley bottom up the neighbouring hillside, but unless it is pushed above the threshold into another valley, the process of development will find its way back. It will not return to the point from which it started, but to some later position on the canalized pathway of change. This represents regulation.

The concept of the chreode is very similar to that of the morphogenetic field, but it makes explicit the dimension of time

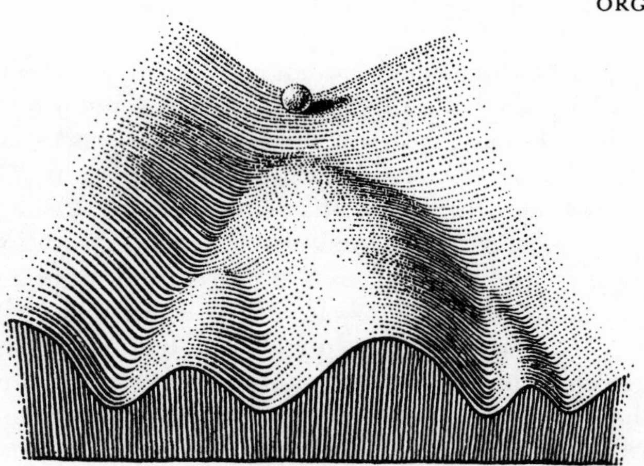

Figure 5 Part of an 'epigenetic landscape,' illustrating the concept of the chreode as a canalized pathway of change. (From Waddington, 1957. Reproduced courtesy of George Allen & Unwin, Ltd.)

which is only implicit within the latter.

Recently, both these concepts have been extensively developed by the mathematician René Thom as part of a comprehensive attempt to create a mathematical theory embracing not only morphogenesis, but also behaviour and language.[31] His main concern is to find an appropriate mathematical formalism for these problems, which have so far resisted mathematical treatment. The final objective is to produce mathematical models which correspond as closely as possible to developmental processes. These models would be topological, qualitative rather than quantitative, and would not depend on any particular scheme of causal explanation: 'One essential feature of our use of local models is that it implies nothing about the "ultimate nature of reality"; even if this is ever revealed by analysis complicated beyond description, only a part of its manifestation, the so-called observables, are finally relevant to the macroscopic description of the system. The phase space of our dynamical model is defined using only these observables and without reference to any more or less chaotic underlying structures.'[32]

The problem with this approach is that it is essentially descriptive; it does little to *explain* morphogenesis. This is indeed the case with all existing organismic theories of morphogenesis. Compare, for example, Waddington's chreode with Driesch's entelechy. Both include the idea that development is guided or canalized in space and time by something which cannot itself be regarded as confined to a particular place and time; both see this as somehow including within itself the end or goal of the developmental process, and both thereby provide a way of thinking about regulation. The main difference between the two is that Driesch tried to say how the process he proposed might actually work, whereas Waddington did not. The concept of the chreode was therefore less open to attack because it remained so vague.[33] In fact, Waddington regarded the concepts of chreodes and morphogenetic fields as 'essentially a descriptive convenience'.[34] Like a number of other organicists, he denied that he was suggesting the operation of anything other than known physical causes.[35] However, not all organicists make this denial; some leave the question open. This explicitly non-committal attitude is illustrated by the following discussion of the morphogenetic field by B.C. Goodwin:

> 'One aspect of the field is that electrical forces can affect it. Other developing and regenerating organisms have also been found to have interesting and significant electrical field patterns, but I would not wish to suggest that the morphogenetic field is essentially electrical. Chemical substances also affect polarity and other spatial aspects of developing organisms; and again I would not wish to draw the conclusion that the morphogenetic field is essentially chemical or biochemical in nature. My belief is that its investigation should proceed on the assumption that it could be any, or all or none of these things; but that, despite agnosticism regarding its material nature, it plays a primary role in the developmental process.'[36]

The openness of this concept makes it the most promising starting point for a detailed organismic theory of morphogenesis. But clearly, if morphogenetic fields are considered to be fully explicable in terms of known physical principles, they represent nothing but an ambiguous terminology superimposed upon a sophisticated version of the mechanistic theory. Only if they are assumed to play a causal role, at present unrecognised by physics, can a testable theory be developed. This possibility is explored in the following chapters.

Notes

1 For an example of the way in which a consideration of results of descriptive research can lead to the formulation of hypotheses, see Crick and Lawrence (1975).
2 For a recent account, see Wolpert (1978).
3 King and Wilson (1975).
4 ibid.
5 MacWilliams and Bonner (1979).
6 Sheldrake (1973).
7 For a recent theoretical discussion of this problem, see Meinhardt (1978).
8 Roberts and Hyams (eds) (1979).
9 Nicolis and Prigogine (1977).
10 In Driesch (1914), p.119.
11 Driesch (1929), p.290.
12 Driesch (1908), Vol. 1, p.203.
13 Driesch (1929), pp.152-4, 293.
14 ibid., p.135, 291.
15 ibid., p.246.
16 ibid., p.103.
17 ibid., p.246.
18 ibid., p.266.
19 ibid., p.262.
20 Eddington (1935), p.302.
21 Eccles (1953).
22 E.g. Walker (1975); Whiteman (1977); Hasted (1978); Lawden (1980).
23 Cf. the concept of 'mnemic causation' discussed by Bertrand Russell (1921).
24 The idea that memory and instinct are two aspects of the same phenomenon has been proposed, among others, by Butler (1878); Semon (1921); and Rignano (1926). However, these authors assumed that the inheritance of memory took place physically, through the germ plasm, which would require a kind of Lamarckian inheritance.
25 Carington (1945).
26 Hardy (1965), p.257.
27 For a discussion of these influences, and an account of the subsequent development of organismic ideas see Haraway (1976). The best early summary of the organismic approach to morphogenesis is by von Bertalanffy (1933).
28 Gurwitsch (1922).
29 For a systematic statement of P. Weiss's ideas, see his *Principles of Development* (1939).

30 Waddington (1957), Chapter 2.
31 Thom (1975a).
32 ibid., pp.6-7.
33 Waddington did not even make explicit the organismic background of his concepts, for the reason explained in the following passage, written towards the end of his career:

> 'Since I am an unaggressive character, and was living in an aggressively anti-metaphysical period, I chose not to expound publicly these philosophical views. An essay I wrote around 1928 on "The Vitalist-Mechanist Controversy and the Process of Abstraction" was never published. Instead I tried to put the Whiteheadian outlook to use in particular experimental situations. So biologists uninterested in metaphysics do not notice what lies behind – though they usually react as though they feel obscurely uneasy.' (Waddington (ed.), 1969, pp.72-81).

34 In Waddington (ed.) (1969), pp.238, 242.
35 E.g. Elsasser (1966, 1975); von Bertalanffy (1971). For a discussion of this 'mechanistic organicism' see Sheldrake (1981).
36 Goodwin (1979), pp.112-113.

3 The Causes of Form

3.1 The problem of form

It is not immediately obvious that form presents any problem at all. The world around us is full of forms; we recognise them in every act of perception. But we easily forget that there is a vast gulf between this aspect of our experience, which we simply take for granted, and the quantitative factors with which physics concerns itself: mass, momentum, energy, temperature, pressure, electric charge, etc.[1]

The relationships between the quantitative factors of physics can be expressed mathematically, and physical changes can be described by means of equations. The construction of these equations is possible because fundamental physical quantities are conserved according to the Principles of Conservation of Mass and Energy, Momentum, Electric Charge, etc.: the total amount of mass and energy, momentum, electric charge, etc. before a given physical change equals the total amount afterwards. But form does not enter into these equations: it is not a vector or scalar quantity, nor is it conserved. For example, if a bunch of flowers is thrown into a furnace and reduced to ashes, the total amount of matter and energy remains the same, but the form of the flowers simply disappears.

Physical quantities can be measured with instruments to a high degree of accuracy. But forms cannot be measured on a quantitative scale, nor do they need to be, even by scientists. A botanist does not measure the difference between two species on the dial of an instrument; nor does an entomologist recognise butterflies by means of a machine, nor an anatomist bones, nor a histologist cells. All these forms are recognised directly. Then specimens of plants are preserved in herbaria, butterflies and bones in cabinets, and cells on microscope slides. As forms they are simply themselves; they cannot be reduced to anything else. The description and classification of forms is in fact the primary concern of many branches of science; even in a physical science such as chemistry, a major objective is the elucidation of the forms of molecules, represented diagrammatically in two-dimensional 'structural formulae' or in three-dimensional models of the 'ball and stick' type.

The forms of all but the simplest systems can only be represented visually, whether by photographs, drawings, diagrams or models. They cannot be represented mathematically. Even the most advanced topological methods are not yet sufficiently developed to be capable of providing mathematical formulae for, say, a giraffe or an oak tree. Some of the new methods being developed by Thom and others may eventually be able to deal with problems such as these, but there are mathematical difficulties not only in practice but in principle.[2]

If the mere description of any but the simplest static forms presents a mathematical problem of appalling complexity, the description of change of form, of morphogenesis, is even more difficult. This is the subject of Thom's 'catastrophe theory', which classifies and describes in general terms the possible types of change of form, or 'catastrophe'. He applies this theory to the problems of morphogenesis by constructing mathematical models in which the end or goal of a morphogenetic process, the final form, is represented by an 'attractor' within a morphogenetic field. He postulates that every object, or physical form, can be represented by such an attractor and that all morphogenesis 'can be described by the disappearance of the attractors representing the initial forms, and their replacement by capture by the attractors representing the final forms'.[3]

In order to develop topological models which correspond to particular morphogenetic processes, formulae are found by a combination of trial and error and inspired guess-work. If a mathematical expression gives too many solutions, restrictions have to be introduced into it; and if a function is too restricted, a more generalised function is used instead. By methods such as these, Thom hopes that it should eventually be possible to develop topological expressions which correspond in detail to actual morphogenetic processes. But even so, these models would probably not enable quantitative predictions to be made. Their main value might lie in drawing attention to formal analogies between different types of morphogenesis.[4]

At first sight, the mathematical formalism of Information Theory may seem preferable to this topological approach. But in fact Information Theory is severely limited in its scope. It was originally developed by telephone engineers in connection with the transmission of messages from a source, through a channel, to a receiver; it was primarily concerned with the question of how the characteristics of a channel influence the amount of information that can be transmitted in a given time. One of the basic results is that in a closed system, no more information can be transmitted to the receiver than was contained in the source, although the form of the information can be changed, for example from the dots and dashes of Morse code to words. The information content of an event is defined not by what has happened, but only with respect to what might have happened instead. For this purpose binary symbols are usually used, and then the information content of a pattern is determined by finding out how many yes or no decisions are needed to specify which particular class of a pattern out of a known number of classes has occurred.

In biology this theory has some relevance to the quantitative study of the transmission of impulses by nerve fibres; to a lesser extent it has a bearing on the transmission of a sequence of bases in the DNA of parents to the DNA of their offspring, although even in such a simple case as this it can be seriously misleading, because in living organisms things happen which do not occur in telephone wires: genes mutate, parts of chromosomes undergo inversions, translocations, etc. But Information Theory is not relevant to biological morphogenesis: it applies only to the transmission of information within closed systems, and it cannot allow for an increase in the content of information during this process.[5] Developing organisms are not closed systems, and their development is epigenetic, i.e. the complexity of form and organization increases. Although mechanistic biologists often speak of 'genetic information', 'positional information', etc. as if these terms had some well-defined meaning, this is an illusion: they borrow only the jargon of Information Theory, and leave its rigour behind.

However, even if impressively detailed mathematical models of morphogenetic processes could be made by whatever method, and even if they gave rise to predictions which agreed with experimental evidence, there would still be the question of what these models corresponded to. Indeed the same question is raised by the correspondence between mathematical models and empirical observations in any branch of science.

One answer is provided by a mathematical mysticism of the Pythagorean type: the universe is seen as dependent upon a fundamental mathematical order which somehow gives rise to all empirical phenomena; this transcendent order is revealed and becomes comprehensible only through the methods of mathematics. Although this attitude is rarely advocated explicitly, it has a strong influence within modern science, and can often be found, more or less thinly disguised, among mathematicians and physicists.

Alternatively, the correspondence can be explained by the tendency of the mind to seek and find order in experience: the ordered structures of mathematics, creations of the human mind, are superimposed onto experience, and those that do not fit are discarded; thus by a process resembling natural selection, those mathematical formulae which fit best are retained. In this view, scientific activity is concerned only with the development and empirical testing of mathematical models of more or less isolated and definable aspects of the world; it cannot lead to any fundamental understanding of reality.

However, in relation to the problem of form, there is a different type of approach which necessitates neither an acceptance of Pythagorean mysticism, nor the abandoning of the possibility of explanation. If the forms of things are to be understood, they need not be explained in terms of *numbers*, but in terms of more fundamental *forms*. Plato considered that the forms in the world of sense-experience were like imperfect reflections of transcendent, archetypal Forms or Ideas. But this doctrine, strongly influenced by the mysticism of the Pythagoreans, failed to explain how the eternal Forms were related to the changing world of phenomena. Aristotle believed this problem could be overcome by regarding the forms of things as immanent, rather than transcendent: specific forms were not only inherent in objects, but actually *caused* them to take up their characteristic forms.

This type of alternative to Pythagorean mysticism has been developed in modern non-mechanistic theories of morphogenesis. In Driesch's system, which was explicitly based on that of Aristotle, the specific forms of living organisms were caused by a non-energetic agency, entelechy. The morphogenetic fields and chreodes of the organicists play a similar role in guiding morphogenetic processes towards specific final forms. But their nature has so far remained obscure.

This obscurity may be due, in part, to the Platonic tendency of much organismic thought,[6] most clearly apparent in A.N. Whitehead's system of philosophy. Whitehead postulated that all actual events involved what he called Eternal Objects; the latter collectively made up the realm of possibility, and included all possible forms; indeed, they strongly resembled Platonic Forms.[7] But clearly, a metaphysical notion of morphogenetic fields as aspects of Platonic Forms or Eternal Objects would be of little value to experimental science. Only if they are regarded as physical entities which have physical effects can they help to provide a scientific understanding of morphogenesis.

The organismic philosophy embraces both biology and physics; hence if morphogenetic fields are assumed to play a causal role in biological morphogenesis, they should also play a causal role in the morphogenesis of simpler systems such as crystals and molecules. Such fields are not recognised in the existing theories of physics. Therefore it is important to consider to what extent these existing theories are capable of explaining the morphogenesis of purely chemical systems. If they are able to provide an adequate explanation, then the idea of morphogenetic fields is gratuitous; but if they are not, the way lies open for a new hypothesis of the causation of form through morphogenetic fields in both biological and non-biological systems.

3.2 Form and energy

In Newtonian physics, all causation was seen in terms of *energy*, the principle of movement and change.

All moving things have energy – the kinetic energy of moving bodies, thermal vibration and electromagnetic radiation – and this energy can cause other things to move. Static things may also have energy, a potential energy which is due to their tendency to move; they are only static because they are restrained by forces which oppose this tendency.

Gravitational attraction was thought to depend on a force which acted at a distance causing bodies to move, or giving them a tendency to move, a potential energy. However, no reason could be given for the existence of this attractive force itself. By contrast, gravitational as well as electromagnetic effects are now explained in terms of *fields*. Whereas the Newtonian forces were supposed to arise in some unexplained way from material bodies and to spread out from them into space, in modern physics the fields are primary: they underlie both material bodies and the space in between them.

This picture is complicated by the fact that there are several different types of field. First, the gravitational field, which in Einstein's General Theory of Relativity is equated with space-time, and considered to be curved in the presence of matter. Secondly, the electromagnetic field, within which electrical charges are localised, and through which electromagnetic radiations propagate as vibrational disturbances. According to the quantum theory, these disturbances are regarded as particle-like photons associated with discrete quanta of energy. Thirdly, in the quantum field theory of matter, subatomic particles are thought of as quanta of excitation of matter fields. Each kind of particle has its own special type of field: a proton is a quantum of the proton-antiproton field, an electron a quantum of the electron-positron field, and so on.

In these theories, physical phenomena are explained by a combination of the concepts of spatial fields and of energy, not in terms of energy alone. Thus although energy can be regarded as the cause of change, the *ordering* of change depends on the spatial structure of the fields. These structures have physical effects, but they are not in themselves a type of energy; they act as 'geometrical' or spatial causes. The radical difference between this idea and the notion of exclusively energetic causation is illustrated in the contrast between Newton's and Einstein's theories of gravitation: for example, according to the former the moon moves around the earth because it is pulled towards it by an attractive force; according to the latter, it does so because the very space in which it moves is curved.

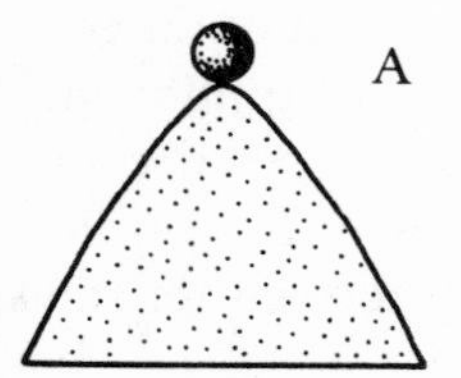

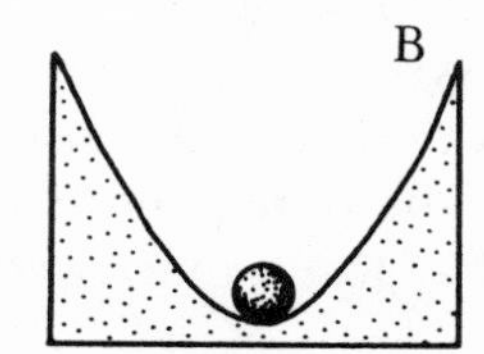

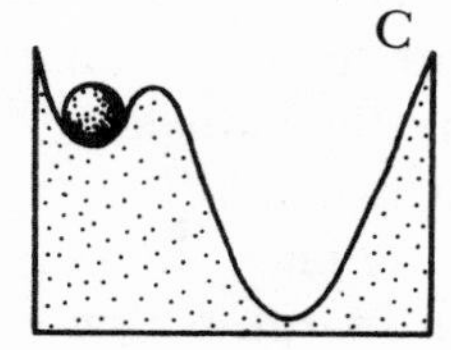

Figure 6 A diagrammatic representation of unstable (A), stable (B) and partially stable (C) states.

The modern understanding of the structure of chemical systems depends on the concepts of quantum mechanics and of electromagnetism; gravitational effects are very small by comparison and can be ignored. The possible ways in which the atoms can combine together are given by the Schrödinger equation of quantum mechanics, which enables the orbitals of electrons to be calculated in terms of probabilities; in the quantum field theory of matter these orbitals can be regarded as structures within the electron-positron field. But since electrons and atomic nuclei are electrically charged, they are also associated with spatial patterns within the electromagnetic field, and hence with potential energies. Not all the possible spatial arrangements of a given number of atoms have the same potential energy, and only the arrangement with the lowest potential energy will be stable for reasons indicated in Fig. 6. If a system is in a state which has a higher energy than possible alternative states, any small displacement (for example due to thermal agitation) will cause it to move into another state (A). If, on the other hand, it is in a state with a lower energy than possible alternatives, after small displacements it will return to this state, which is consequently stable (B). A system may also exist temporarily in a state which is not the most stable so long as it is not displaced above the level of a 'barrier' (C); when this happens it will move into a stabler, lower-energy state.

These energetic considerations determine which is the most stable state of a chemical structure, but they do not account for its spatial characteristics, which in Fig. 6 are represented by the slopes down which the ball rolls, and which act as barriers confining it. These depend on spatial patterns given by the fields of matter and electromagnetism.

According to the second law of thermodynamics, spontaneous processes within a closed system tend towards a state of equilibrium; as they do so, initial differences in temperature, pressure etc. between different parts of the system tend to disappear. In technical language, the entropy of a closed macroscopic system either stays the same or increases.

The significance of this law is often exaggerated in popular accounts; in particular, the term entropy is treated as if it was synonymous with 'disorder'. Then the increasing complexity of organisation which occurs in the evolution and development of living organisms appears to contradict the principle of increasing entropy. This confusion arises from a misunderstanding of the limitations of the science of thermodynamics. First, it applies only to closed systems, whereas living organisms are open systems, exchanging matter and energy with their environment. Secondly, it deals only with the inter-relations between heat and other forms of energy: it is relevant to the energetic factors which affect chemical and biological structures, but does not account for the existence of these structures in the first place. And thirdly, the technical definition of entropy bears little relation to any non-technical conception of disorder; in particular, it is not concerned with the type of order inherent in the specific structures of chemical and biological systems. According to the third law of thermodynamics, at absolute zero the entropies of all pure crystalline solids are zero. They are perfectly 'ordered' from a thermodynamic point of view because there is no disorder due to thermal agitation. But all are equally ordered: there is no difference in entropy between a simple salt crystal and a crystal of an extremely complex organic macromolecule such as haemoglobin. It follows that the greater structural complexity of the latter is not measurable in terms of entropy.

The contrast between 'order' in the sense of chemical or biological structure, and thermodynamic 'order' owing to inequalities of temperature, etc. in a large system consisting of countless atoms and molecules is illustrated by the process of crystallization. If a solution of a salt is placed in a dish inside a cold enclosure, the salt crystallizes as the solution cools. Initially, its constituent ions are distributed at random within the solution, but as crystallization takes place they become ordered with great regularity within the crystals, and the crystals themselves develop into macroscopically symmetrical structures. From a morphological point of view, there has been a considerable increase in order; but from a thermodynamic point of view there has been a decrease in 'order', an increase in entropy, owing to the equalization of temperature between the solution and its surroundings, and to the release of heat during the process of crystallization, leading to a greater thermal agitation of the solvent molecules.

Similarly, when an animal embryo grows and develops, there is an increase in entropy of the thermodynamic system consisting of the embryo and the environment from which it takes its food and to which it releases heat and excretory products. The second law of thermodynamics serves to emphasise this dependence of living organisms on external sources of energy, but it does nothing to explain their specific forms.

In the most general terms, form and energy bear an inverse relationship to each other: energy is the principle of change, but a form or structure can only exist as long as it has a certain stability and resistance to change. This opposition is clearly apparent in the relationship between the states of matter and temperature. Under sufficiently cool conditions, substances exist in crystalline forms in which the arrangements of the molecules show a high degree of regularity and order. As the temperature is raised, at a certain point the thermal energy causes the crystalline form to disintegrate; the solid melts. In the liquid state the molecules arrange themselves in transient patterns which constantly shift and change. The forces between the molecules create a surface tension which imparts simple forms to the liquid as a whole, as in spherical drops. With a further rise in temperature the liquid vaporizes; in the gaseous state the molecules are isolated and behave more or less independently of each other. At higher temperatures still, the molecules themselves disintegrate into atoms, and at yet higher temperatures even the atoms break up to give a mixed gas of electrons and atomic nuclei, a plasma.

When this sequence is viewed in reverse, progressively more complex and organized structures appear as the temperature is reduced, the most stable ones first and the least stable ones last. As a plasma cools, appropriate numbers of electrons congregate around atomic nuclei in their appropriate orbitals. At lower temperatures atoms come together into molecules. Then as the gas condenses into droplets, supra-molecular forces come into play. Finally when the liquid crystallizes a high degree of supra-molecular order is established.

These forms appear spontaneously. They cannot be explained in terms of external energy, except negatively in the sense that they can come into being and persist only below a certain temperature. They can be explained in terms of internal energy only to the extent that out of all the possible structural arrangements, only the one with the lowest potential energy will be stable; this is therefore the structure that will spontaneously tend to be taken up.

3.3 The prediction of chemical structures

Quantum mechanics is able to describe in detail the electronic orbitals and the energy states of the simplest of all chemical systems, the hydrogen atom. With more complicated atoms and with even the simplest chemical molecules its methods are no longer so precise; the complexity of the calculations becomes formidable, and only approximate methods can be used. For complex molecules and crystals detailed calculations are impossible, at least in practice. The structures of the molecules and the atomic arrangements within crystals can be found out empirically, by chemical and crystallographic methods; these structures may indeed be more or less predictable by chemists and crystallographers on the basis of empirical laws. But this is a very different matter from providing a fundamental explanation of chemical structures by means of the Schrödinger wave equation.

It is important to realize this severe limitation of quantum mechanics. Certainly it helps to provide a qualitative or semi-quantitative understanding of chemical bonds and of certain aspects of crystals, such as the difference between insulators and electrical conductors. But it has not enabled the forms and properties of even simple molecules and crystals to be predicted from first principles. The situation is even worse with regard to the liquid state, of which there is still no satisfactory quantitative account. And it is illusory to imagine that quantum mechanics in any detailed or rigorous way explains the forms and properties of the very complex molecules and macro-molecular aggregates studied by biochemists and molecular biologists, not to mention the vastly greater complexity of form and properties of even the simplest living cell.

So widespread is the assumption that chemistry provides a firm foundation for the mechanistic understanding of life, it is perhaps necessary to emphasize on what very slender foundations of physical theory chemistry itself rests. In the words of Linus Pauling:

> 'We may believe the theoretical physicist who tells us that all the properties of substances should be calculable by known methods – the solution of the Schrödinger equation. In fact, however, we have seen that during the 30 years since the Schrödinger equation was discovered only a few accurate non-empirical quantum-mechanical calculations of the properties of substances in which the chemist is interested have been made. The chemist must still rely upon experiment for most of his information about the properties of substances'.[8]

Although a further 20 years have passed since this passage was published, and although there have been important improvements in the approximate methods of calculation available to quantum chemists, the situation remains essentially the same today.

Nevertheless, it may be argued that the detailed calculations could be carried out in principle. But even assuming for the purpose of argument that these calculations could indeed be performed, it cannot be known in advance that they will be *correct*, that is to say agree with empirical observations. So at present there is no evidence for the conventional assumption that complex chemical and biological structures can be fully explained in terms of existing physical theory.

The reasons for the difficulty, if not impossibility, of predicting the form of a complex chemical structure on the basis of the properties of its constituent atoms can perhaps be understood more clearly by means of a simple illustration. Consider elementary building blocks which can be added to each other one at a time either endways or sideways (Fig. 7). With two building blocks there are $2^2=4$ possible combinations; with three, $2^3=8$; with four, $2^4=16$; with five $2^5=32$; with ten, $2^{10}=1,024$ with twenty, $2^{20}=1,048,576$; with thirty, $2^{30}=1,073,741,824$; and so on. The number of possibilities soon becomes enormous.

In a chemical system, the different possible arrangements of

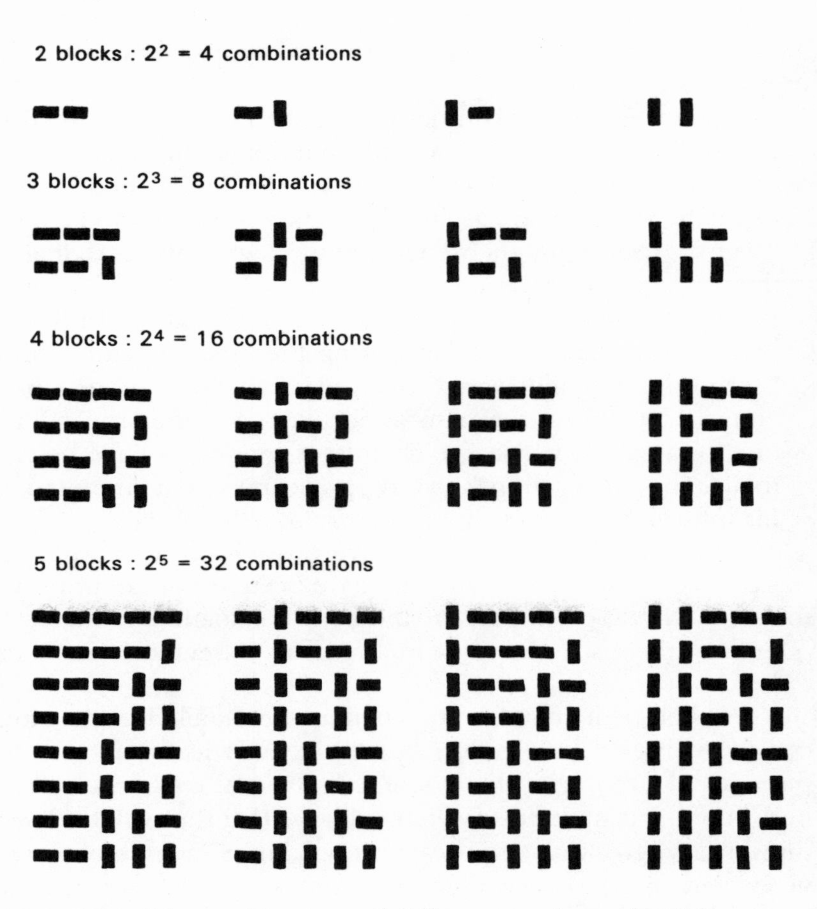

Figure 7 Possible combinations of different numbers of building blocks capable of being joined together either endways or sideways.

atoms have different potential energies owing to the electrical and other interactions between them; the system will spontaneously tend to take up the structure with the minimum potential energy. In a simple system with only a few possible structures, one may have a distinctly lower energy than the others; in Fig. 8A this is represented by the minimum at the bottom of the 'potential well'; other less stable possibilities are represented by local minima on the side of the 'well'. In systems of increasing complexity, the number of possible structures increases (Fig. 8 B,C,D); as it does so the chance of there being a *unique* minimum-energy structure

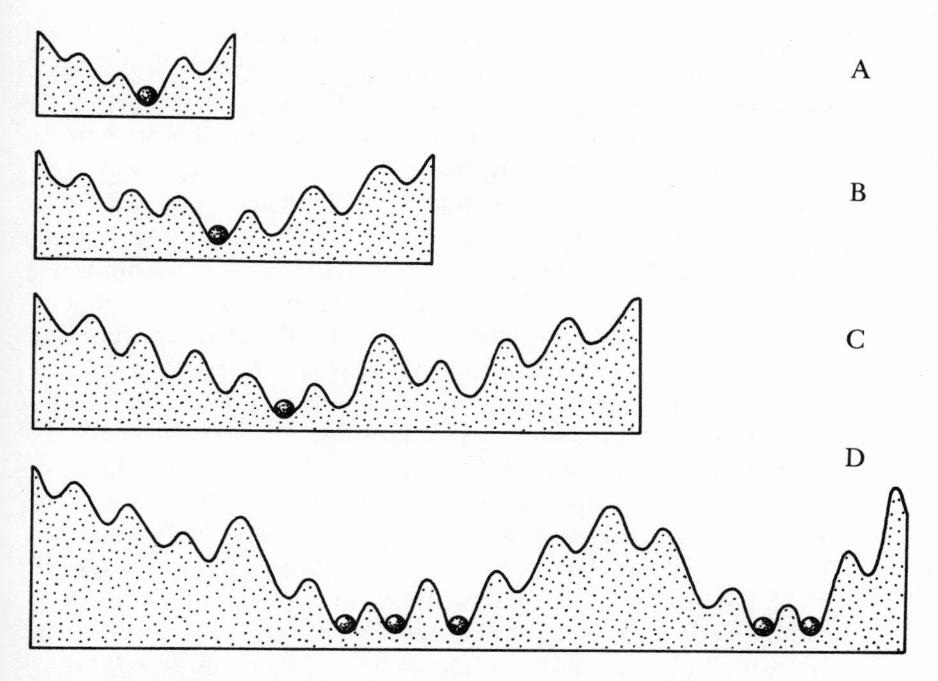

Figure 8 A diagrammatic representation of the possible structures of systems of increasing complexity. In A there is a unique minimum-energy structure, but in D several different possible structures are equally stable.

seems likely to diminish. In the situation represented by Fig. 8 D several different structures would be equally stable from an energetic point of view. If the system were found to take up any of these possible structures at random, or if it oscillated between them, then there would be no problem. But if it invariably took up only one of these structures, this would indicate that some factor other than energy somehow determined that this particular structure was realized rather than the other possibilities. No such factor is at present recognized by physics.

Although chemists, crystallographers and molecular biologists cannot carry out the detailed calculations necessary to predict the minimum-energy structure or structures of a system *a priori*, they are able to use various approximate methods in combination with empirical data on the structures of similar substances. In general, these calculations do not permit unique structures to be predicted (except for the simplest of systems), but only a range of possible structures with more or less equal minimum energies. Thus these approximate results appear to support the idea that energetic considerations are insufficient to account for the unique structure of a complex chemical system. But this conclusion can always be avoided by re-asserting that the unique stable structure must have a lower energy than any other possible structure. This assertion could never be falsified because in practice only approximate methods of calculation can be used; the unique structure actually realized could therefore always be attributed to subtle energetic effects which eluded calculation.

The following discussion of Pauling's illustrates the situation with regard to the structure of inorganic crystals:

> 'Simple ionic substances such as the alkali halogenides have little choice of structure; and a very few relatively stable ionic arrangements corresponding to the formula M^+X^- exist, and the various factors that influence the stability of the crystal are pitted against one another, with no one factor necessarily finding clear expression in the decision between the sodium chloride and the caesium chloride arrangements. For a complex substance, such as mica, $KAl_3Si_3O_{10}(OH)_2$, or zunyite, $Al_{13}Si_5O_{20}(OH)_{18}Cl$, on the other hand, many conceivable structures differing only slightly in nature and stability can be suggested, and it might be expected that the most stable of these possible structures, the one actually assumed by the substance, will reflect in its various features the different factors which are of significance in determining the structure of ionic crystals. It has been found possible to formulate a set of rules about the stability of complex ionic crystals . . . These rules were obtained in part by induction from the structures known in 1928, and in part by deduction from the equations of crystal energy. They are not rigorous in their derivation nor universal in their application, but they have been found useful as a criterion for the probable correctness of reported structures for complex crystals and an aid to X-ray investigation of crystals by making possible the suggestion of reasonable structures for experimental test.'[9]

The range of possible structures becomes much greater in organic chemistry, especially in the case of macromolecules such as

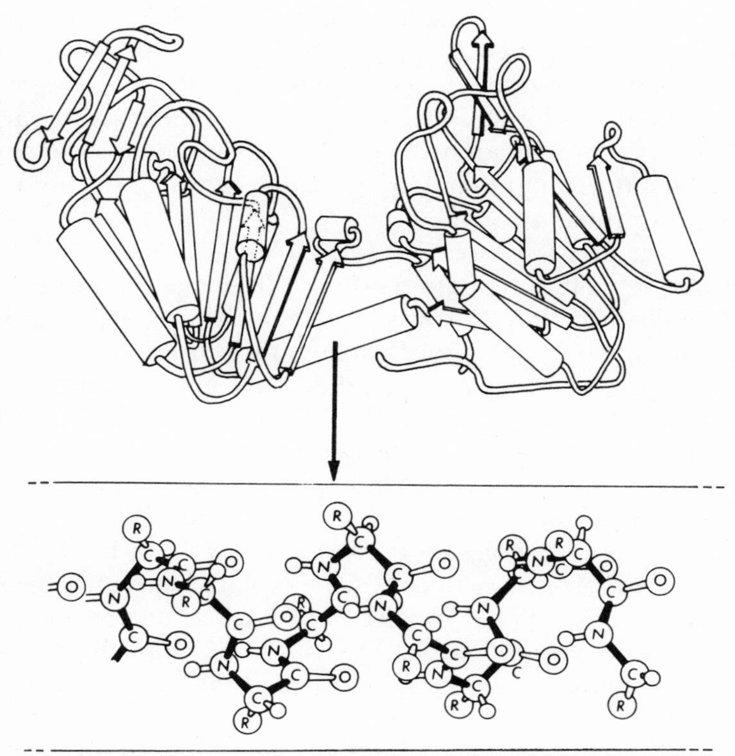

Figure 9 Above: The structure of the enzyme phosphoglycerate kinase, isolated from horse muscle. α – helices are represented by cylinders and ß – strands by arrows.
Below: The structure of an α – helical region in more detail. (After Banks *et al.*, 1979).

proteins, the polypeptide chains of which twist, turn and fold into complicated three-dimensional forms (Fig. 9). There is good evidence that under conditions in which a given type of protein molecule is stable, it folds up into a unique structure. In numerous experimental studies, proteins have been made to unfold to varying degrees by changing their chemical environment; they have then been found to fold up again into their normal structure when they are replaced in appropriate conditions; in spite of starting from different initial states and following different 'pathways' of folding, they reach the same structural end-point.[10]

This stable end-point is likely to be a minimum-energy structure.

But this does not prove that it is the *only* possible structure with a minimum energy; there may be many other possible structures with the same minimum energy. Indeed, calculations designed to predict the three-dimensional structure of proteins, using various methods of approximation, invariably give far too many solutions. In the literature on protein folding, this is known as the 'multiple-minimum problem'.[11]

There are persuasive reasons for thinking that the protein itself does not 'test' all these minima until it finds the right one:

> 'If the chain explored all possible configurations at random by rotations about the various single bonds of the structure, it would take too long to reach the native configuration. For example, if the individual residues of an unfolded polypeptide chain can exist in only two states, which is a gross underestimate, then the number of possible randomly generated conformations is 10^{45} for a chain of 150 amino acid residues (although, of course, most of these would probably be sterically impossible ones). If each conformation could be explored with a frequency of a molecular rotation (10^{12}sec.$^{-1}$), which is an overestimate, it would take approximately 10^{26} years to examine all possible conformations. Since the synthesis and folding of a protein chain such as that of ribonuclease or lysozyme can be accomplished in about 2 minutes, it is clear that all conformations are not traversed in the folding process. Instead, it appears to us that, in response to local interactions, the peptide chain is directed along a variety of possible low-energy pathways (relatively small in number), possibly passing through unique intermediate states, towards the conformation of lowest free energy.' (C.B. Anfinsen and H.A. Scheraga.[12])

But not only may the folding process be 'directed' along certain pathways, it may also be directed towards one particular conformation of minimum energy, rather than any other possible conformations with the same minimum energy.

This discussion leads to the general conclusion that the existing theories of physics may well be incapable of explaining the unique structures of complex molecules and crystals; they permit a range of possible minimum-energy structures to be suggested, but there is no evidence that they can account for the fact that one rather than another of these possible structures is realized. It is therefore conceivable that some factor other than energy 'selects' between these possibilities and thus determines the specific structure taken up by the system.[13] The hypothesis that will now be developed is based on the idea that this 'selection' is brought about by a new type of causation, at present unrecognized by physics, through the agency of morphogenetic fields.

3.4 Formative causation

The hypothesis of *formative causation* proposes that morphogenetic fields play a causal role in the development and maintenance of the forms of systems at all levels of complexity. In this context, the word 'form' is taken to include not only the shape of the outer surface or boundary of a system, but also its internal structure. This suggested causation of form by morphogenetic fields is called formative causation in order to distinguish it from the energetic type of causation with which physics already deals so thoroughly.[14] For although morphogenetic fields can only bring about their effects in conjunction with energetic processes, they are not in themselves energetic.

The idea of non-energetic formative causation is easier to grasp with the help of an architectural analogy. In order to construct a house, bricks and other building materials are necessary; so are the builders who put the materials into place; and so is the architectural plan which determines the form of the house. The same builders doing the same total amount of work using the same quantity of building materials could produce a house of different form on the basis of a different plan. Thus the plan can be regarded as a *cause* of the specific form of the house, although of course it is not the only cause: it could never be realized without the building materials and the activity of the builders. Similarly, a specific morphogenetic field is a cause of the specific form taken up by a system, although it cannot act without suitable 'building blocks' and without the energy necessary to move them into place.

This analogy is not intended to suggest that the causative role of morphogenetic fields depends on conscious design, but only to emphasize that not all causation need be energetic, even though all processes of change involve energy. The plan of a house is not in itself a type of energy. Even when it is drawn on paper, or finally realized in the form of the house, it does not weigh anything or have any energy of its own. If the paper is burnt or the house is demolished, there is no measurable change in the total amount of mass and energy; the plan simply vanishes. Likewise, according to the hypothesis of formative causation, morphogenetic fields are not in themselves energetic; but nevertheless they play a causal role in determining the forms of the systems with which they are associated. For if a system were associated with a different morphogenetic field, it would develop differently.[15] This hypothesis is empirically testable in cases where the morphogenetic fields acting on systems can be altered (Sections 5.6, 7.4, 7.6, 11.2, and 11.4 below).

Morphogenetic fields can be regarded as analogous to the known fields of physics in that they are capable of ordering physical changes, even though they themselves cannot be observed directly. Gravitational and electromagnetic fields are spatial structures which are invisible, intangible, inaudible, tasteless and odourless; they are detectable only through their respective gravitational and electromagnetic effects. In order to account for the fact that physical systems influence each other at a distance without any apparent material connection between them, these hypothetical fields are endowed with the property of traversing empty space, or even actually constituting it. In one sense, they are non-material; but in another sense they are aspects of matter because they can only be known through their effects on material systems. In effect, the scientific definition of matter has simply been widened to take them into account. Similarly, morphogenetic fields are spatial structures detectable only through their morphogenetic effects on material systems; they too can be regarded as aspects of matter if the definition of matter is widened still further to include them.

Although in the preceding Sections only the morphogenesis of biological and complex chemical systems has been discussed, the hypothesis of formative causation will be assumed to apply to biological and physical systems at all levels of complexity. Since each kind of system has its own characteristic form, each must have a specific kind of morphogenetic field: thus there must be one kind of morphogenetic field for protons; another for nitrogen atoms; another for water molecules; another for sodium chloride crystals; another for the muscle cells of earthworms; another for the kidneys of sheep; another for elephants; another for beech trees; and so on.

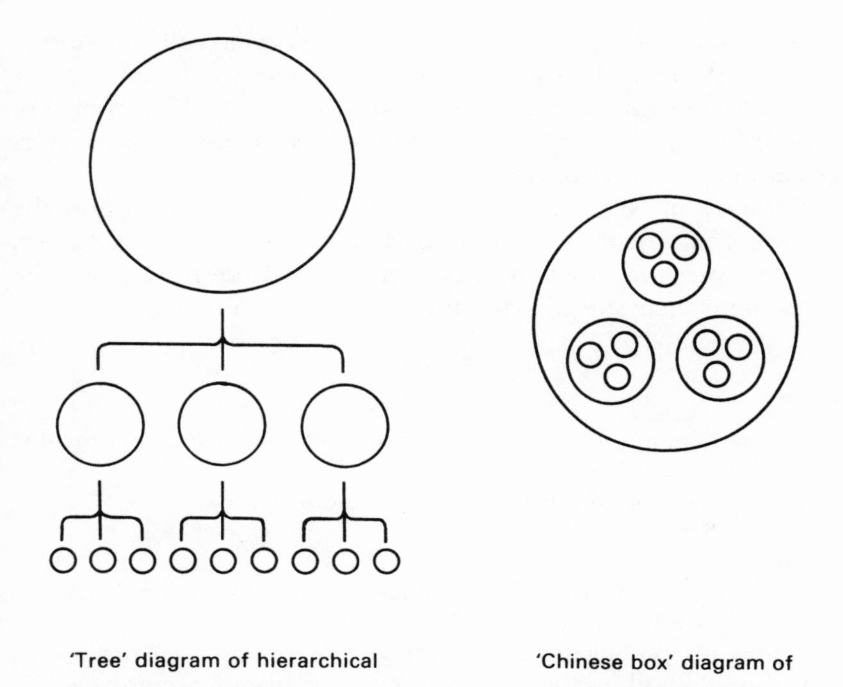

Figure 10 Alternative ways of representing a simple hierarchical system.

According to the organismic theory, systems or 'organisms' are hierarchically organized at all levels of complexity.[16] In the present discussion these systems will be referred to as *morphic units*. The adjective morphic (from the Greek root morphē=form) emphasizes the aspect of structure, and the word unit the unity or wholeness of the system. In this sense, chemical and biological systems are composed of hierarchies of morphic units: a crystal, for example, contains molecules, which contain atoms, which contain sub-atomic particles. Crystals, molecules, atoms, and sub-atomic particles are morphic units, as are animals and plants, organs, tissues, cells and organelles. Simple examples of this hierarchical type of organization can be visualized diagrammatically either as a 'tree' or as a series of 'chinese boxes' (Fig. 10).

A higher-level morphic unit must somehow co-ordinate the

arrangement of the parts of which it is composed. It will be assumed to do so through the influence of its morphogenetic field on the morphogenetic fields of lower-level morphic units. Thus morphogenetic fields, like morphic units themselves, are essentially hierarchical in their organization.

The way in which morphogenetic fields might act upon the systems under their influence is discussed in the following chapter; and the question of where they themselves come from and what gives them their specific structure is discussed in Chapter 5.

Notes

1 An excellent introduction to the problem of organic form is provided by Sinnott (1963).

2 For a discussion of this problem, see Thom (1975 a).

3 ibid., p.320.

4 Thom (1975 b).

5 For a discussion of the limited relevance of Information Theory to biology, see Waddington (1975), pp. 209-230.

6 Numerous examples of the combination of aspects of the organismic philosophy with explicitly neo-Platonic speculation are provided by Ruyer (1974) in his account of a small neo-gnostic group in the United States, whose members include a number of prominent scientists.

7 See Emmet (1966).

8 Pauling (1960), p.220.

9 ibid., p.543.

10 Anfinsen and Scheraga (1975).

11 For a recent review see Némethy and Scheraga (1977).

12 Anfinsen and Scheraga (1975).

13 Cf. Elsasser's (1975) 'principle of finite classes'.

14 This distinction between formative causation and energetic causation resembles Aristotle's distinction between 'formal causes' and 'efficient causes'. However, the hypothesis of formative causation developed in the following chapters differs radically from Aristotle's theory, which presupposed eternally given forms.

15 From a theoretical point of view, the causal role of morphogenetic fields can be analysed in terms of 'counterfactual conditionals'. For a discussion of the latter, see for example Mackie (1974).

16 Arthur Koestler has suggested the use of the term *holon* for such 'self-regulating open systems which display both the autonomous properties of wholes and the dependent properties of parts' (in Koestler and

Smythies (eds) (1969), pp. 210-211). This term is wider in its application than the term morphic unit – it includes for example linguistic and social structures – but it represents a very similar concept.

4 Morphogenetic Fields

4.1 Morphogenetic germs

Morphogenesis does not take place in a vacuum. It can only begin from an already organized system which serves as a *morphogenetic germ*. During morphogenesis a new higher-level morphic unit comes into being around this germ, under the influence of a specific morphogenetic field. So how does this field become associated with the morphogenetic germ to start with?

The answer may be that just as the association of material systems with gravitational fields depends on their mass, and with electromagnetic fields on their electrical charge, so the association of systems with morphogenetic fields depends on their form. Hence a morphogenetic germ becomes surrounded by a particular morphogenetic field because of its characteristic form.

The morphogenetic germ is a part of the system-to-be. Therefore part of the system's morphogenetic field corresponds to it. However the rest of the field is not yet 'occupied' or 'filled out'; it contains the *virtual form* of the final system, which is actualized only when all its material parts have taken up their appropriate places. The morphogenetic field is then in coincidence with the actual form of the system.

These processes are represented diagrammatically in Fig. 11 A. The stippled areas indicate the virtual form and the solid lines the actual form of the system. The morphogenetic field can be thought of as a structure surrounding or embedding the morphogenetic germ, and containing the virtual final form; this field then orders events within its range of influence in such a way that the virtual form is actualized. In the absence of the morphic units which constitute the parts of the final system, this field is undetectable; it reveals itself only through its ordering effects on these parts when they come within its influence. A rough analogy is provided by the 'lines of force' in the magnetic field around a magnet; these spatial structures are revealed when particles capable of being magnetized, such as iron filings, are introduced into the vicinity. Nevertheless, the magnetic field can be considered to exist even when the iron filings are absent; likewise, the morphogenetic field around a morphogenetic germ exists as a spatial structure even though it has not yet been actualized in the final form of the system. However, morphogenetic fields differ radically from electromagnetic fields in that the latter depend on the *actual* state of the system – on the distribution and movement of charged particles – whereas morphogenetic fields correspond to the *potential* state of a developing system and are already present before it takes up its final form.[1]

In Fig. 11 A, there are several intermediate stages between the morphogenetic germ and the final form. The final form could also be reached by a different morphogenetic pathway (Fig. 11 B), but if a particular pathway is usually followed, this can be regarded as a 'canalized pathway of change', or chreode (cf. Fig. 5).

If the developing system is damaged by the removal of a part of it, it may still be able to reach the final form (Fig. 11 C). This represents regulation.

After the final form is actualized, the continued association between the morphogenetic field and the system whose form corresponds to it will tend to stabilize the latter. Any deviations of the system away from this form will tend to be corrected as the system is attracted back towards it. And if part of the system is removed, the final form will tend to be actualized again (Fig. 11 D). This represents regeneration.

The type of morphogenesis shown in Fig. 11 is essentially *aggregative*: previously separate morphic units come together into a higher-level morphic unit. Another type of morphogenesis is possible when the morphic unit which serves as the morphogenetic germ is already part of a different higher-level morphic unit. The influence of the new morphogenetic field leads to a *transformation* in which the form of the original higher-level morphic unit is replaced by the form of the new one. Most types of chemical morphogenesis are aggregative, whereas biological morphogenesis usually involves a combination of transformative and aggregative processes. Exam-

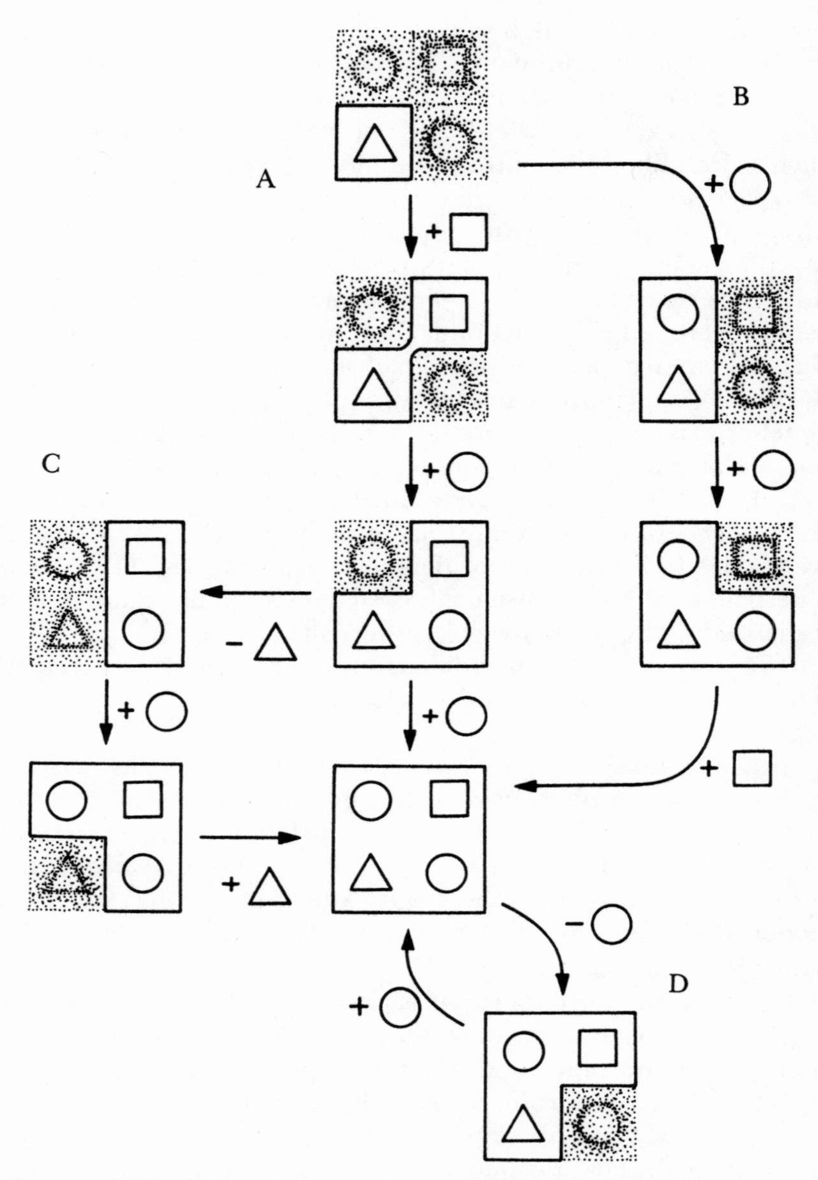

Figure 11 Diagrammatic representation of the development of a system from a morphogenetic germ (triangle) by the normal chreode, A. An alternative morphogenetic pathway is represented by B, regulation by C, and regeneration by D. The virtual form within the morphogenetic field is indicated by the stippled area.

ples are considered in the following Sections.

4.2 Chemical morphogenesis

Aggregative morphogeneses occur progressively in inorganic systems as the temperature is reduced: as a plasma cools, sub-atomic particles aggregate into atoms; at lower temperatures atoms aggregate into molecules; then molecules condense into liquids; and finally liquids crystallize.

In the plasma state, the naked atomic nuclei can be regarded as the morphogenetic germs of atoms; they are associated with the atomic morphogenetic fields, which contain the virtual orbitals of electrons. In one sense these orbitals do not exist, but in another sense they have a reality which is revealed in the cooling plasma as they are actualized by the capture of electrons.

Electrons which have been captured within atomic orbitals may be displaced from them again through the influence of external energy, or by entering a virtual orbital of lower potential energy. In the latter case, they lose a discrete quantum of energy which is radiated as a photon. In atoms with many electrons, each orbital can contain only two electrons (with opposing spins); thus in a cooling plasma the virtual orbitals with the lowest potential energies fill up with electrons first, then the orbitals with the next lowest energies, and so on until the complete atomic form has been actualized around the morphogenetic germ of the nucleus.

Atoms are in turn the morphogenetic germs of molecules, and small molecules the germs of larger molecules. Chemical reactions involve either the aggregation of atoms and molecules into larger molecules – for example in the formation of polymers – or the fragmentation of molecules into smaller ones, or into atoms and ions, which may then aggregate with others, for example in combustion: under the influence of external energy, molecules fragment into atoms and ions which then combine with those of oxygen to form small, simple molecules like H_2O and CO_2. These chemical changes involve the actualization of virtual forms associated with the atoms or molecules which act as morphogenetic germs.

The idea that molecules have virtual forms before they are actualized is illustrated particularly clearly by the familiar fact that entirely new compounds can first be 'designed' on the basis of empirically determined principles of chemical combination and then actually synthesized by organic chemists. These laboratory syntheses are carried out step by step; in each step a particular molecular form serves as the morphogenetic germ for the next virtual form to be synthesized, ending up with the form of the entirely new molecule.

If it seems rather artificial to think of chemical reactions as morphogenetic processes, it should be remembered that much of the effect of catalysts, both inorganic and organic, depends on their morphology. For example, enzymes, the specific catalysts of the numerous reactions of biochemistry, provide surfaces, grooves, notches or basins into which the reacting molecules fit with a specificity which is often compared to that of a lock and key. The catalytic effect of enzymes depends to a large extent on the way in which they hold reactant molecules in the appropriate relative positions for reaction to occur. (In free solution the chance collisions of the molecules occur in all possible orientations, most of which are inappropriate.)

The details of chemical morphogeneses are vague, partly because of their great rapidity, partly because the intermediate forms may be highly unstable, and also because the ultimate changes consist of probabilistic quantum jumps of electrons between the orbitals which constitute the chemical bonds. The virtual form of the molecule-to-be is outlined in the morphogenetic field associated with the atomic or molecular morphogenetic germ; when the other atom or molecule approaches in an appropriate orientation, the form of the product molecule is actualized by means of quantum jumps of electrons into orbitals which previously existed only as virtual forms; at the same time energy is released, usually as thermal motion. The role of the morphogenetic field in this process is, as it were, energetically passive, but morphologically active; it creates virtual structures which are then actualized as lower-level morphic units 'slot' or 'snap' into them, releasing energy as they do so.

Any given type of atom or molecule can take part in many different types of chemical reaction, and it is therefore the potential germ of many different morphogenetic fields. These could be thought of as possibilities 'hovering' around it. However, it may not take on its role as the germ of a particular morphogenetic field until an appropriate reagent atom or molecule approaches it, perhaps owing to specific electromagnetic or other effects of the latter upon it.

The morphogenesis of crystals differs from that of atoms and molecules in that a particular pattern of atomic or molecular arrangement is repeated indefinitely. The morphogenetic germ is provided by this pattern itself. It is well known that the addition of 'seeds' or 'nuclei' of the appropriate type of crystal greatly accelerates the crystallization of supercooled liquids or supersaturated solutions. In the absence of these seeds or nuclei, morphogenetic germs of the crystal come into being only when the atoms or molecules take up their appropriate relative positions by chance, owing to thermal agitation. Once the germ is present, the virtual forms of repetitions of the lattice structure given by the morphogenetic field extend outwards from the surfaces of the growing crystal. Appropriate free atoms or molecules which approach these surfaces are captured and 'slot' into position; again thermal energy is released as they do so.

The seeding or nucleation of supercooled liquids or supersaturated solutions can also be carried out, although less effectively, with small fragments of unrelated substances; for example, chemists often scratch the sides of test tubes to seed solutions with fragments of glass. These fragments provide surfaces which facilitate the appropriate relative positioning of the atoms or molecules which constitute the true morphogenetic germ of the crystal. In their morphogenetic effect these seeds resemble the catalysts of chemical reactions.

All the types of chemical morphogenesis considered so far are essentially aggregative. Transformations are much less common in non-living systems. Most crystals, for example, cannot undergo transformations into other crystalline forms; they can be melted or dissolved, and then their constituents may take part in other processes of crystallisation; but this is a disaggregation followed by new types of aggregation. Chemical reactions likewise involve disaggregative and aggregative changes. There are, however, important examples of molecular transformation, such as the folding of proteins, and the reversible changes of shape which occur when certain enzymes bind to the molecules whose reaction they catalyse.[2]

The fact that proteins fold up far more rapidly than would be expected if they 'found' their final form by a 'random search' indicates that their folding follows particular pathways, or a limited number of pathways (Section 3.3). These 'canalized pathways of change' can be regarded as chreodes. For the folding process to begin, according to the ideas developed in Section 4.1 above, a morphogenetic germ must be present, and this germ must already have the characteristic three-dimensional structure that it has in the final form of the protein. The existence of such morphogenetic starting points has in fact already been suggested in the literature on protein folding:

> 'The extreme rapidity of the refolding makes it essential that the process takes place along a limited number of pathways, even when the statistics are severely restricted by the kinds of stereochemical ground rules that are implicit in a so-called Ramachandran plot. It becomes necessary to postulate the existence of a limited number of allowable initiating events in the folding process. Such events, generally referred to as nucleations, are most likely to occur in parts of the polypeptide chain that can participate in conformational equilibria between random and cooperatively stabilized arrangements . . . Furthermore it is important to stress that the amino acid sequences of polypeptide chains designed to be the fabric of protein molecules only make functional sense when they are in the three-dimensional arrangement that characterizes them in the native protein molecule. It seems reasonable to suggest that portions of a protein chain that can serve as nucleation sites for folding will be those that can 'flicker' in and out of the conformation that they occupy in the final protein, and that they will form a relatively rigid structure stabilized by a set of cooperative interactions.' (C.B. Anfinsen.[3])

Such 'nucleation sites' would act as morphogenetic germs through their association with the morphogenetic field of the protein, which would then canalize the pathway of folding towards the characteristic final form.

4.3 Morphogenetic fields as 'probability structures'

The orbitals of electrons around an atomic nucleus can be regarded as structures within the morphogenetic field of the atom. These orbitals can be described by solutions of the Schrödinger equation. However, according to quantum mechanics, the precise orbits of electrons cannot be specified, but only the probabilities of finding electrons at particular points; the orbitals are regarded as probability distributions in space. Within the context of the hypothesis of formative causation, this result suggests that just as these structures in the morphogenetic fields of atoms must be thought of as spatial probability distributions, so morphogenetic fields in general are not precisely defined, but are given by probability distributions.[4] It will be assumed that this is in fact the case, and the structures of morphogenetic fields will henceforth be referred to as *probability structures*.[5] An explanation for the probabilistic nature of these fields will be suggested in Section 5.4.

The action of the morphogenetic field of a morphic unit on the morphogenetic fields of its parts, which are morphic units at lower levels (Section 3.4), can be thought of in terms of the influence of this higher-level probability structure on lower-level probability structures; the higher-level field modifies the probability structures of the lower-level fields. Consequently, during morphogenesis, the higher-level field modifies the probability of probabilistic events in the lower-level morphic units under its influence.[6]

In the case of free atoms, electronic events take place with the probabilities given by the unmodified probability structures of the atomic morphogenetic fields. But when the atoms come under the influence of the higher-level morphogenetic field of a molecule, these probabilities are modified in such a way that the probability of events leading towards the actualization of the final form are enhanced, while the probability of other events is diminished. Thus the morphogenetic fields of molecules *restrict* the possible number of atomic configurations which would be expected on the basis of calculations which start from the probability structures of free atoms. And this is what is found in fact; in the case of protein folding, for example, the rapidity of the process indicates that the system does not 'explore' the countless configurations in which the atoms could conceivably be arranged (Section 3.3).

Similarly, the morphogenetic fields of crystals restrict the large number of possible arrangements which would be permitted by the probability structures of their constituent molecules; hence one particular pattern of molecular arrangement is taken up as the substance crystallizes, rather than any of the other conceivable structures.

Thus the morphogenetic fields of crystals and molecules are probability structures in the same sense as the electronic orbitals in the morphogenetic fields of atoms are probability structures. This conclusion agrees with the conventional assumption that there is no difference in kind between the description of simple atomic systems by quantum mechanics and a potential quantum mechanical description of more complex forms. But unlike the hypothesis of formative causation, the conventional theory seeks to explain complex systems from the bottom up, as it were, in terms of the quantum mechanical properties of atoms.

The difference between these two approaches can be seen more clearly in an historical context. Quantum theory itself was primarily elaborated in connection with the properties of simple systems such as hydrogen atoms. As time went on, new fundamental principles were introduced in order to account for empirical observations such as those on the fine structure of the spectra of light emitted by atoms. The original quantum numbers characterizing the discrete electronic orbitals were supplemented by another set referring to angular momentum, and then yet more referring to 'spin'. The latter is considered to be an irreducible property of particles, just as electric charge is, and has its own conservation law. In nuclear particle physics, still more irreducible factors, such as 'strangeness' and 'charm', together with further conservation laws have been introduced more or less *ad hoc* in order to account for observations not explicable in terms of the already accepted quantum factors. Moreover, the discovery of large numbers of new sub-atomic particles has led to the postulation of an ever-increasing number of new kinds of matter field.

When so many new physical principles and physical fields have been introduced in order to account for the properties of atoms and sub-atomic particles, the conventional assumption that no new physical principles or fields come into play at levels of organization above that of the atom seems remarkably arbitrary. It is in fact little more than a relic of nineteenth century atomism; now that atoms are no longer regarded as ultimate and indivisible, its original theoretical justification has vanished. From the point of view of the hypothesis of formative causation, although the existing body of quantum theory, developed in connection with the properties of atoms and sub-atomic particles, sheds much light on the nature of morphogenetic fields, it cannot be extrapolated to describe the morphogenetic fields of more complex systems. There is no reason why the morphogenetic fields of atoms should be considered to have a privileged position in the order of nature; they are simply the fields of morphic units at one particular level of complexity.

4.4 Probabilistic processes in biological morphogenesis

There are many examples of physical processes whose spatial outcomes are probabilistic. In general, changes involving the breaking of a symmetry or homogeneity are indeterminate; examples occur in the phase transitions between the gaseous and liquid states, and the liquid and solid states. If, for instance, a spherical balloon filled with vapour is cooled below saturation point in the absence of external gradients of temperature and gravity, the liquid will start by condensing on the walls, but the final distribution of the liquid will be unpredictable, and almost never spherically symmetrical.[7] Thermodynamically, the relative amounts of liquid and vapour can be forseen, but their spatial distribution cannot. In the crystallization of a substance under uniform conditions, the spatial distribution and the numbers and sizes of the crystals cannot be predicted; in other words, if the same process were to be repeated under identical conditions, each time the spatial outcome would differ in detail.

The forms of crystals themselves, although they exhibit a definite symmetry, may be indeterminate; a familiar example is provided by snow-flakes, which come in a myriad of different shapes.[8]

In the 'dissipative structures' of macroscopic physical and chemical systems far from thermodynamic equilibrium, random fluctuations can give rise to spatial patterns, for example convection cells in a heated liquid, or coloured bands in solutions in which the Zhabotinski reaction is proceeding. The mathematical descriptions of such cases of 'order through fluctuations' by the methods of non-equilibrium thermodynamics show striking analogies to phase transitions.[9]

These examples of spatial indeterminism are drawn from quite simple physical and chemical processes. In living cells, the physico-chemical systems are far more complex than any encountered in the inorganic realm, and include many potentially indeterminate phase transitions and non-equilibrium thermodynamical processes. In the protoplasm there are crystalline, liquid and lipid phases in dynamic inter-relation; then there are numerous types of macro-molecule which can come together into crystalline or quasi-crystalline aggregates; lipid membranes, which as 'liquid crystals' hover on the borderline between the liquid and solid states, as do the colloidal sols and gels; electrical potentials across membranes which fluctuate unpredictably; and 'compartments', containing different concentrations of inorganic ions and other substances, separated by membranes across which these substances move probabilistically.[10] With such complexity, the number of energetically possible patterns of change must be enormous, and there is thus a vast scope for the operation of morphogenetic fields through the imposition of patterns on these probabilistic processes.

This is not to say that *all* form in living organisms is determined by formative causation. Some patterns may come about through random processes. Others may be fully explicable in terms of minimum-energy configurations: for instance, the spherical shape of free-floating egg cells (e.g. those of sea urchins) may be fully explicable in terms of the surface tension of the cell membrane. However, the very limited success of simple physical explanations of biological forms[11] suggests that *most* aspects of biological morphogenesis are determined by morphogenetic fields. It should be re-emphasized that these fields do not act alone, but together with the energetic and chemical causes studied by biophysicists and biochemists.

One example of the way morphogenetic fields could operate within cells is provided by the positioning of microtubules, tiny rod-like structures formed by the spontaneous aggregation of protein sub-units. Microtubules play an important role as microscopic 'scaffolds' within both animal and plant cells: they guide and orientate processes such as cell division (the spindle fibres in mitosis and meiosis are made up of microtubules), and the patterned deposition of cell wall material in differentiating plant cells; they also serve as intra-cellular 'skeletons', maintaining particular cellular shapes, as in radiolarians.[12] Now if the spatial distribution of microtubules is responsible for the patterning of many different sorts of process and structure within cells, then what controls the spatial distribution of the microtubules? If other patterns of organization are responsible,[13] the problem is simply pushed back one stage: what controls these patterns of organization themselves? But the problem cannot be pushed back indefinitely, because development is epigenetic, that is to say it involves an increase in spatial diversity and organization which cannot be accounted for in terms of preceding patterns or structures; sooner or later something else has to account for the emergence of the pattern in which the microtubules aggregate.

On the present hypothesis, this pattern is due to the action of specific morphogenetic fields. These fields greatly increase the probability of aggregation of microtubules in appropriate dispositions either directly, or indirectly through the establishment of a preceding pattern of organization. Obviously, the patterning activity of the fields depends on the presence of a supersaturated solution of microtubule sub-units within the cell, and on appropriate physico-chemical conditions for their aggregation: these are necessary conditions for the formation of microtubules, but they are not in themselves sufficient to account for the pattern in which the microtubules appear.

The objection might be raised that the suggested action of formative causation in patterning probabilistic processes within cells is impossible because it would lead to a local violation of the second law of thermodynamics. But this objection is not valid. The second law of thermodynamics refers only to assemblies of very large numbers of particles, and not to processes on a microscopic scale. Moreover it applies only to closed systems: a region of a cell is not a closed system, nor of course are living organisms in general.

In living organisms, as in the chemical realm, the morphogenetic fields are hierarchically organized: those of organelles – for example the cell nucleus, the mitochondria and chloroplasts – act by ordering physico-chemical processes within them; these fields are subject to the higher-level fields of cells; the fields of cells to those of tissues; those of tissues to those of organs; and of organs to the morphogenetic field of the organism as a whole. At each level the fields work by ordering processes which would otherwise be indeterminate. For example, at the cellular level the morphogenetic field orders the crystallization of microtubules and other processes which are necessary for the co-ordination of cell division. But the planes in which the cells divide may be indeterminate in the absence of a higher-level field: for instance, in plant wound-calluses the cells proliferate more or less randomly to produce a chaotic mass.[14] Within an organized tissue, on the other hand, one of the functions of the tissue's morphogenetic field may be to impose a pattern on the planes of cell division, and thus control the way in which the tissue as a whole grows. Then the development of tissues may itself be inherently indeterminate in many respects, as revealed when they are artificially isolated and grown in tissue culture;[15] under normal conditions this indeterminacy is restricted by the higher-level field of the organ. Indeed at each level in biological systems, as in chemical systems, the morphic units in isolation behave more indeterminately than they do when they are part of a higher-level morphic unit. The higher-level morphogenetic field restricts and patterns their intrinsic indeterminism.

4.5 Morphogenetic germs in biological systems

At the cellular level, the germs for morphogenetic transformations must be lower-level morphic units within the cells which are present both at the beginning and at the end of the process of cellular differentiation. The possible morphogenetic germs of these transformations are not immediately obvious: they could be organelles, macromolecular aggregates, cytoplasmic or membranous structures, or the cell nuclei. In many cases nuclei might well play this role. But since so many different types of differentiated cell can be produced in the same organism, if the nuclei are to act as morphogenetic germs they must be capable of taking on different patterns of organization in the different cell types: the differentiation of a cell must be preceded by a differentiation of its nucleus, owing to changes in its membrane, or in the arrangement of its chromosomes, or in the associations between proteins and nucleic acids within the chromosomes, or in the nucleoli, or in other components. Such changes could be brought about directly or indirectly through the action of the higher-level morphogenetic field of the differentiating tissue. There is indeed considerable evidence that many types of cellular differentiation are preceded by nuclear changes. The suggestion advanced here diverges from the usual interpretation of these changes in regarding their significance as not simply chemical, owing to the production of special types of messenger RNA, but, in addition, as morphogenetic: the modified nuclei might serve as germs with which the specific morphogenetic fields of differentiated cells become associated.[16]

There is at least one process of cellular morphogenesis in which the nucleus cannot be the morphogenetic germ: in nuclear division. It loses its identity as a separate structure when the nuclear membrane breaks down and disappears.[17] The doubled, highly coiled chromosomes become aligned in the equatorial region of the mitotic spindle and a complete set then moves to each spindle pole. Then new nuclear membranes develop around each set of chromosomes to form the daughter nuclei. The morphogenetic germs for these processes must be extra-nuclear structures or organelles, and there must be two of them.[18]

The development of tissues and organs usually involves both transformative and aggregative changes. In these morphogeneses, the morphogenetic germs must be cells or groups of cells which are present both as part of the final form and also at the beginning of the morphogenetic process; they cannot be those specialized cells which appear only after the process has begun. Thus the morphogenetic germs seem likely to be relatively unspecialized cells which undergo little change. In higher plants such cells are present, for example, in the apical zones of the meristems or growing points.[19] In shoots, the flowering stimulus transforms the meristems in such a way that they give rise to flowers rather than leaves and other vegetative structures; the apical zones, suitably modified by the flowering stimulus, could be the morphogenetic germs for this transformation. In animal embryos, many 'organizing centres' which play a key role in the development of tissues and organs have already been identified by experimental embryologists; one example is the apical ectodermal ridge at the tip of developing limb buds.[20] These 'organizing centres' may well be the germs with which the major morphogenetic fields become associated.

Although both in the chemical and biological realms, morphogenetic germs can be suggested, if not actually identified, much remains obscure, especially the reason for the particular form of each morphogenetic field and for the way in which it becomes associated with its germ. The consideration of these problems in the following chapter leads to a more complete hypothesis of formative causation which, although surprising and unfamiliar, is perhaps less difficult to understand.

Notes

1 The identification of morphogenetic fields with electromagnetic fields is responsible for much of the confusion inherent in H.S. Burr's theory of electrodynamic 'Life Fields'. Burr (1972) cites indisputable evidence that living organisms are associated with electromagnetic fields, which change as the organisms change, but then goes on to argue that these fields control morphogenesis by acting as 'blueprints' for development, which is a very different matter.

2 For a review of the literature on confomational changes in proteins in solution, see Williams (1979).

3 Anfinsen (1973), p.228.

4 For a general discussion of probabilistic causality, see Suppes (1970).

5 Cf. Sir Karl Popper's concept of probability or propensity fields (Popper, 1967; Popper and Eccles, 1977).

6 This suggestion might fit in with the approach to quantum physics advocated by Bohm (1969, 1980) and Hiley (1980).

7 This and other instances of what R. Thom calls 'generalised catastrophes' are discussed in Chapter 6 of his *Structural Stability and Morphogenesis*.

8 Bentley and Humphreys (1962).

9 See Nicolis and Prigogine (1977). A different but related approach to these problems is outlined by Haken (1977).

10 Stevens (1977).

11 Sir D'Arcy Thompson in his classic essay *On Growth and Form* (1942) suggested that many aspects of biological morphogenesis could be explained in terms of physical forces: for example, the plane of cell division in terms of surface tension, which would tend to give a minimum surface area. But there are so many exceptions that these simple interpretations have met with very little success. For a discussion of Thompson's theories see Medawar (1968).

12 For recent accounts of the properties and functions of microtubules see Dustin (1978) and Roberts and Hyams (eds) (1979).

13 One suggestion is that the smooth endoplasmic reticulum plays a role in the transport of microtubule sub-units to the regions in which they aggregate (Burgess and Northcote, 1968). The existence of 'nucleating elements' which may or may not be bound together in 'microtubule

organizing centres' has also been suggested (J.B. Tucker, in Roberts and Hyams (eds), 1979).

14 Street and Henshaw (1965).

15 For examples, see Willmer (1970).

16 In some instances the nuclei are destroyed in the final stages of differentiation (e.g. xylem vessels in plants, red blood cells in mammals). In these cases the nuclei could act as morphogenetic germs for the differentiation process up to the point at which they were still intact; then the final stages of differentiation could proceed purely mechanistically by straightforward chemical processes not subject to morphogenetic ordering, through the release of hydrolytic enzymes.

17 In some algae, e.g. *Oedogonium*, the nuclear membrane remains intact during mitosis; this is probably an evolutionarily primitive feature (Pickett-Heaps, 1975).

18 In animals the centrioles may appear to be likely candidates for this role, but higher plants have no centrioles. In both cases 'microtubule organizing centres' may well be responsible for the development of the spindle poles; the centrioles may merely be 'passengers' assured of equal partitioning into daughter cells by association with these centres (Pickett-Heaps, 1969). (The centrioles serve as organizing centres, or morphogenetic germs, for the development of cilia and flagella.)

19 Clowes (1961).

20 Wolpert (1978).

5 The Influence of Past Forms

5.1 The constancy and repetition of forms

Time after time when atoms come into existence electrons fill the same orbitals around the nuclei; atoms repeatedly combine to give the same molecular forms; again and again molecules crystallize into the same spatial patterns; seeds of a given species give rise year after year to plants of the same appearance; generation after generation, spiders spin the same types of web. Forms come into being repeatedly, and each time each form is more or less the same. On this fact depends our ability to recognize, identify and name things.

This constancy and repetition would present no problem if all forms were uniquely determined by changeless physical laws or principles. This assumption is implicit in the conventional theory of the causation of form. These fundamental physical principles are taken to be temporally prior to the actual forms of things: theoretically, the way in which a newly synthesized chemical will crystallize should be calculable before its crystals appear for the first time; likewise, the effects of a given mutation in the DNA of an animal or plant on the form of the organism should be predictable in advance. Of course in practice such calculations have never been made; this comfortable assumption is untested, and most probably untestable.

By contrast, according to the hypothesis of formative causation, the forms of complex chemical and biological systems are not uniquely determined by the known laws of physics. These laws permit a range of possibilities between which formative causes select. The constancy and repetition of forms is explained by the repeated association of the same type of morphogenetic field with a given type of physico-chemical system. But then what determines the particular form of the morphogenetic field?

One possible answer is that morphogenetic fields are eternal. They are simply given, and are not explicable in terms of anything else. Thus even before this planet appeared, there already existed in a latent state the morphogenetic fields of all the chemicals, crystals, animals and plants that have ever occurred on the earth, or that will ever come into being in the future.

This answer is essentially Platonic, or even Aristotelian in so far as Aristotle believed in the eternal fixity of specific forms. It differs from the conventional physical theory in that these forms would not be predictable in terms of energetic causation; but it agrees with it in taking for granted that behind all empirical phenomena lie pre-existing principles of order.

The other possible answer is radically different. Chemical and biological forms are repeated not because they are determined by changeless laws or eternal Forms, but because of a *causal influence from previous similar forms*. This influence would require an action across space *and time* unlike any known type of physical action.

On this view, the unique form taken up by a system would not be physically determined in advance of its first appearance. Nevertheless it would be repeated because the form of the first system would itself determine the form taken up by subsequent similar systems. Imagine, for instance, that out of several different possible forms, P, Q, R, S. . . . , all of which are equally probable from an energetic point of view, a system happens to take up form R on the first occasion. Then on subsequent occasions similar systems will also take up form R because of a trans-spatial and trans-temporal influence from the first such system.

In this case, what determines the form on the first occasion? No scientific answer can be given: the question concerns unique and energetically indeterminate events which, *ex hypothesi*, once they have happened are unrepeatable because they themselves influence all subsequent similar events. Science can deal only with regularities, with things that are repeatable. The initial choice of a particular form could be ascribed to chance; or to a creativity inherent in matter; or to a transcendent creative agency. But there is no way in which these different possibilities could be distinguished from each other by experiment. A decision between them could be made only on metaphysical grounds. This question is discussed briefly in the final chapter of this book; but for present purposes it does not matter which of these possibilities is preferred. The hypothesis of formative causation is concerned only with the *repetition* of forms, and not with the reasons for their appearance in the first place.

This new way of thinking is unfamiliar, and it leads into uncharted territory. But only by exploring it does there seem to be any hope of arriving at a new scientific understanding of form and organization in general, and of living organisms in particular. The alternative to going on would be to return to the starting point; the choice would once again be narrowed to that between a simple mechanistic faith and a metaphysical organicism.

In the following discussion, it is proposed that this hypothetical trans-spatial and trans-temporal influence passes through morphogenetic fields and is an essential feature of formative causation.

5.2 The general possibility of trans-temporal causal connections

Although the hypothesis of formative causation proposes a new kind of trans-temporal, or diachronic, causal connection which has not so far been recognized by science, the possibility of 'action at a distance' in time has already been considered in general terms by several philosophers. There seems to be no *a priori* reason for excluding it. J.L. Mackie, for example, wrote as follows:

> 'While we are happiest about contiguous cause-effect relations, and find "action at a distance" over either a spatial or temporal gap puzzling, we do not rule it out. Our ordinary concept of causation does not absolutely require contiguity; it is not part of our idea of causation in a way that would make "C caused E over a spatial, or temporal, or both spatial and temporal, gap, without intermediate links" a contradiction in terms.'[1]

Moreover, from the point of view of the philosophy of science, there is nothing to prevent the consideration of new kinds of causal connection:

> 'Scientific theory in general does not presuppose any particular mode of causal connection between events, but only that it is possible to find laws and hypotheses, expressed in terms of some model, which satisfy the criteria of intelligibility, confirmation, and falsifiability. The mode of causal connection in

each case is shown by the model, and changes with fundamental changes of model.' (M.B. Hesse.[2])

However, although the new kind of causal connection proposed in the hypothesis of formative causation seems to be possible in principle, the plausibility of this hypothesis can be assessed only after predictions deduced from it have been tested empirically.

5.3 Morphic resonance

The idea of a process whereby the forms of previous systems influence the morphogenesis of subsequent similar systems is difficult to express in terms of existing concepts. The only way to proceed is by means of analogy.

The physical analogy which seems most appropriate is that of *resonance*. Energetic resonance occurs when a system is acted on by an alternating force which coincides with its natural frequency of vibration. Examples include the 'sympathetic' vibration of stretched strings in response to appropriate sound waves; the tuning of radio sets to the frequency of radio waves given out by transmitters; the absorption of light waves of particular frequencies by atoms and molecules, resulting in their characteristic absorption spectra; and the response of electrons and atomic nuclei in the presence of magnetic fields to electromagnetic radiation in Electronic Spin Resonance and Nuclear Magnetic Resonance. Common to all these types of resonance is the principle of selectivity: out of a mixture of vibrations, however complicated, the systems respond only to those of particular frequencies.

A 'resonant' effect of form upon form across space and time would resemble energetic resonance in its selectivity, but it could not be accounted for in terms of any of the known types of resonance, nor would it involve a transmission of energy. In order to distinguish it from energetic resonance, this process will be called *morphic resonance*.

Morphic resonance is analogous to energetic resonance in a further respect: it takes place between vibrating systems. Atoms, molecules, crystals, organelles, cells, tissues, organs and organisms are all made up of parts in ceaseless oscillation, and all have their own characteristic patterns of vibration and internal rhythm; the morphic units are dynamic, not static.[3] But whereas energetic resonance depends only on the specificity of response to particular frequencies, to 'one-dimensional' stimuli,[4] morphic resonance depends on three-dimensional patterns of vibration. What is being suggested here is that by morphic resonance the form of a system, including its characteristic internal structure and vibrational frequencies, becomes *present* to a subsequent system with a similar form; the spatio-temporal pattern of the former *superimposes* itself on the latter.

Morphic resonance takes place through morphogenetic fields and indeed gives rise to their characteristic structures. Not only does a specific morphogenetic field influence the form of a system (as discussed in the previous chapter), but also the form of this system influences the morphogenetic field and through it becomes present to subsequent similar systems.

5.4 The influence of the past

Morphic resonance is non-energetic, and morphogenetic fields themselves are neither a type of mass nor energy. Therefore there seems to be no *a priori* reason why it should obey the laws that have been found to apply to the movement of bodies, particles and waves. In particular, it need not necessarily be attentuated by either spatial or temporal separation between similar systems; it could be just as effective over ten thousand miles as over a yard, and over a century as an hour.

The assumption that morphic resonance is indeed unattentuated by time and space will be adopted as a provisional working hypothesis, on the ground of simplicity.

It will also be assumed on the ground of simplicity that morphic resonance takes place only from the *past*; that only morphic units which have already actually existed are able to exert a morphic influence in the present. The notion that *future* systems, which do not yet exist, might be able to exert a causal influence 'backwards' in time may perhaps be logically conceivable;[5] but only if there were persuasive empirical evidence for a physical influence from future morphic units would it become necessary to take this possibility seriously.[6]

However, assuming that morphic resonance occurs only from past morphic units, and that it is not attenuated by the lapse of time

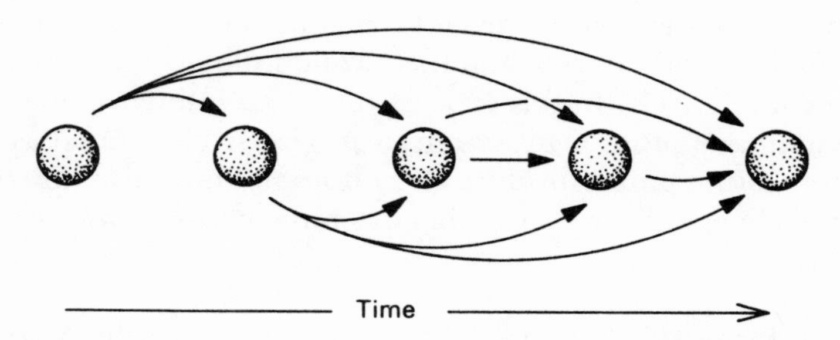

Figure 12 Diagram illustrating the cumulative influence of past systems on subsequent similar systems by morphic resonance.

or by distance, how might it take place? The process can be visualized with the help of several different metaphors. The morphic influence of a past system might become present to a subsequent similar system by passing 'beyond' space-time and then 're-entering' wherever and whenever a similar pattern of vibration appeared. Or it might be connected through other 'dimensions'. Or it might go through a space-time 'tunnel' to emerge unchanged in the presence of a subsequent similar system. Or the morphic influence of past systems might simply be present everywhere. However, these different ways of thinking about morphic resonance would probably not be distinguishable from each other experimentally. All would have the same consequence: the forms of past systems would automatically become present to subsequent similar systems.

An immediate implication of this hypothesis is that a given system could be influenced by *all* past systems with a similar form and pattern of vibration. *Ex hypothesi* the influence of these past systems is not attenuated by temporal or spatial separation. Nevertheless, the ability of past systems to influence subsequent systems could be weakened or exhausted by action; they could have only a limited potential influence which is expended in morphic resonance. This possibility is discussed in Section 5.5 below. But first consider the possibility that their potential action is not reduced in this way, with the consequence that the forms of all past systems influence all subsequent similar systems (Fig.12). This postulate has several important consequences:

(i) The first system with a given form influences the second such system, then both the first and the second influence the third, and so on cumulatively. In this process the direct influence of a given system on any subsequent system is progressively diluted as time goes on; although its absolute effect does not diminish, its *relative* effect declines as the total number of similar past systems increases (Fig.12).

(ii) The forms of even the simplest chemical morphic units are variable: sub-atomic particles are in ceaseless vibratory motion, and atoms and molecules are subject to deformation by mechanical collision and by electrical and magnetic fields. Biological morphic units are still more variable; even if cells and organisms have the same genetic constitution and develop under the same conditions they are unlikely to be identical in every respect.

By morphic resonance, the forms of all similar past systems become present to a subsequent system of similar form. Even assuming that differences in absolute size are adjusted for (see Section 6.3 below), many of these forms will differ from each other in detail. Hence they will not coincide with each other exactly when they are superimposed by morphic resonance. The result will be a process of *automatic averaging* whereby those features which most past systems have in common will be reinforced. However this 'average' form will not be sharply defined within the morphogenetic field, but surrounded by a 'blur' owing to the effect of less common variants.

This process can be visualized more easily by analogy with 'composite photographs' made by superimposing the photographic images of different individuals. As a result of this superimposition the common features are reinforced; but because of the differences between the individual images, the 'average' photographs are not sharply defined (Figs 13 and 14).

(iii) The automatic averaging of past forms will result in a spatial probability distribution within the morphogenetic field, or in other words a *probability structure* (cf. Section 4.3).

The probability structure of a morphogenetic field determines the probable state of a given system under its influence in accordance with the *actual* states of all similar systems in the past;

Figure 13 Photographic portraits of three sisters in full face and profile with the corresponding composites. These pictures are by Francis Galton, who invented the technique of composite photography over a century ago. (From Pearson, 1924. Reproduced by courtesy of the Cambridge University Press).

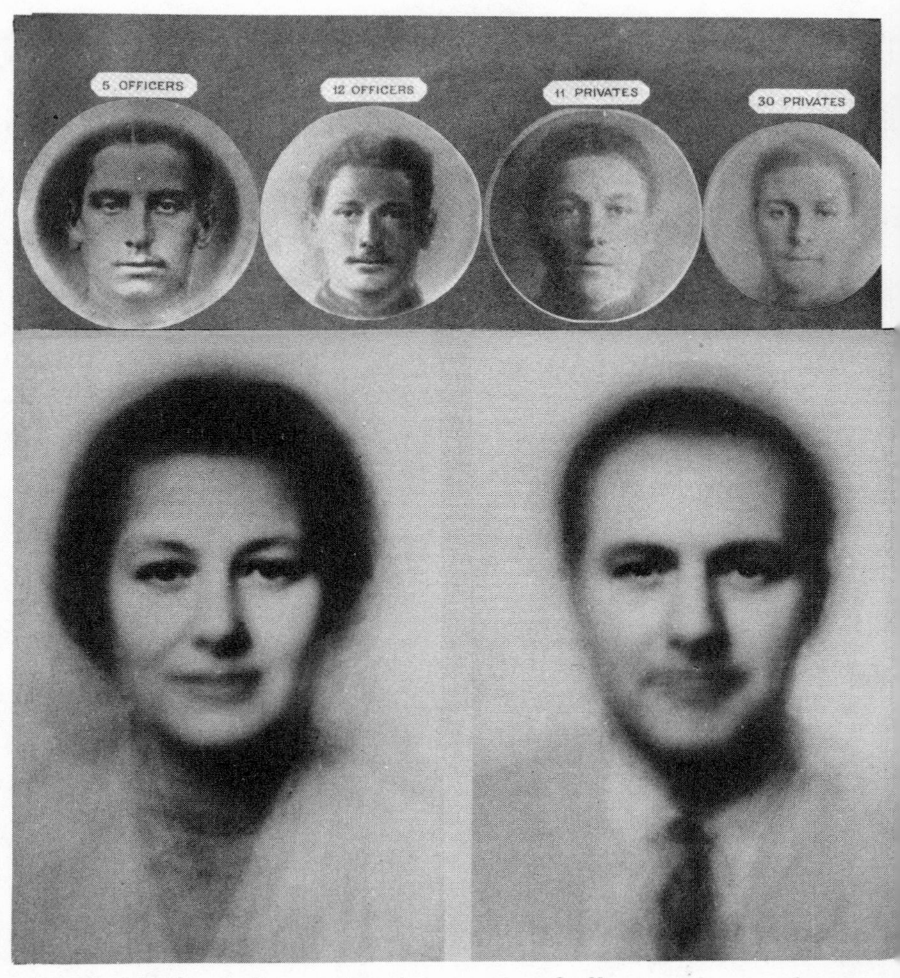

Figure 14 Above: Composite photographs of officers and men of the Royal Engineers, by Francis Galton. (From Pearson, 1924. Reproduced by courtesy of the Cambridge University Press).
Below: Composite photographs of 30 female and 45 male members of the staff of the John Innes Institute, Norwich. (Reproduced by courtesy of the John Innes Institute).

the most probable form the system will take up is that which has occurred most frequently already.

(iv) In the early stages of a form's history, the morphogenetic field will be relatively ill-defined and significantly influenced by individual variants. But as time goes on, the cumulative influence of countless previous systems will confer an ever-increasing stability on the field; the more probable the average type becomes, the more likely that it will be repeated in the future.

To put it in a different way: at first the basin of attraction of the morphogenetic field will be relatively shallow, but it will become progressively deeper as the number of systems contributing to morphic resonance increases. Or to use yet another metaphor, through repetition the form will get into a rut and the more often it is repeated, the deeper will this rut become.

(v) The amount of influence a given system has on subsequent similar systems seems likely to depend on the length of time it survives: one that continues to exist for a year may have more effect than one that disintegrates after a second. Thus the automatic averaging may be 'weighted' in favour of long-lasting previous forms.

(vi) At the beginning of a morphogenetic process, the morphogenetic germ comes into morphic resonance with similar past systems which are part of higher-level morphic units: it thus becomes associated with the morphogenetic field of the higher-level morphic unit (Section 4.1). Let the morphogenetic germ be represented by morphic unit F and the final form towards which the system is attracted by D-E-F-G-H. Let the intermediate stages in the morphogenesis be as shown in Fig. 15. Now not only will the morphogenetic germ and the intermediate stages enter into morphic resonance with the *final* form of previous similar systems, but the intermediate stages will also enter into morphic resonance with similar intermediate stages E-F, D-E-F, etc. in previous similar morphogeneses. Thus these stages will be stabilized by morphic resonance, resulting in a chreode. The more frequently this particular pathway of morphogenesis is followed, the more will this chreode be reinforced. In terms of the 'epigenetic landscape' model (Fig. 5), the valley of the chreode will be deepened the more often development passes along it.

5.5 Implications of an attenuated morphic resonance

The discussion in the preceding Section was based on the assumption that the morphic influence of a given system is not exhausted in its action on subsequent similar systems, although its *relative* effect is diluted as the number of similar systems increases. The alternative possibility that this influence is somehow exhausted will now be considered.

If such an exhaustion takes place, only if the rate of exhaustion were very fast would it be detectable. Consider first the extreme case in which the influence of a system is expended by morphic

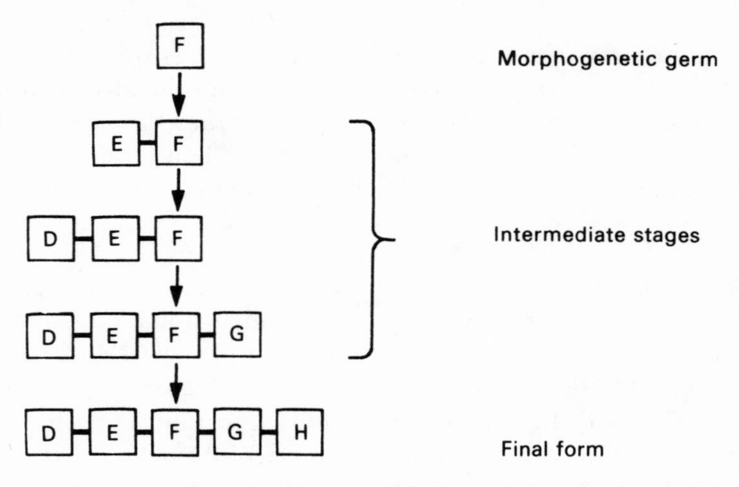

Figure 15 Diagram representing stages in the aggregative morphogenesis of the morphic unit D-E-F-G-H from the morphogenetic germ F.

resonance with only one subsequent system. If the number of similar systems increases with time, then most of them will not be influenced by morphic resonance from previous similar systems (Fig. 16A). They will consequently be free to take up different forms by 'chance' or 'creativity'; the forms of these systems may therefore be very variable.

Next consider the case in which each system can influence two subsequent systems. In the situation represented in Fig. 16B most but not all of the subsequent forms would be stabilized by morphic resonance. If each system influenced three subsequent systems, all

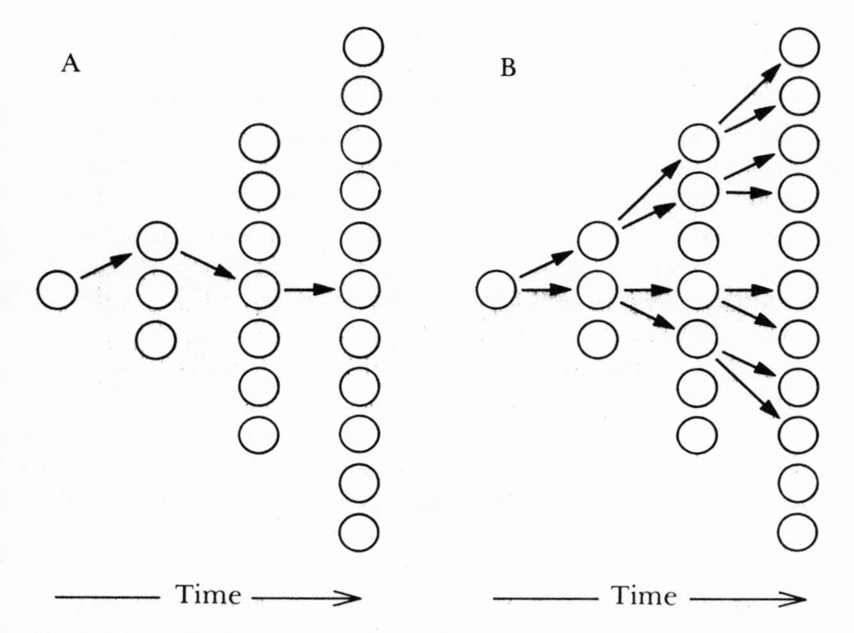

Figure 16 Diagram illustrating situations in which the influence of previous systems is exhausted by morphic resonance with only one subsequent system (A) and two subsequent systems (B).

would be stabilized; an instability of form would appear only if the number of subsequent similar systems increased more rapidly still, as in a population explosion. And if each system influenced many subsequent systems, this low but finite rate of exhaustion of morphic influence would be practically undetectable.

For the sake of simplicity, it will be assumed that the morphic influence of systems on subsequent similar systems is *not* exhausted at all; but it should be borne in mind that this assumption is provisional. The question could eventually be investigated empirically, at least to the extent of distinguishing between a rapid rate of exhaustion of morphic influence on the one hand, and a slow or zero rate on the other.

5.6 A possible experimental test

According to the conventional theory, the unique forms of chemical and biological systems should be predictable in terms of the principles of quantum mechanics, electromagnetism, energetic causation, etc. before they come into being for the first time. By contrast, according to the hypothesis of formative causation, unique forms will not be predictable in advance, but only a range of possible forms. Thus, in principle, the failure of the conventional theory to give rise to unique predictions would provide evidence against it and in favour of the hypothesis of formative causation. But in practice this failure could never be conclusive: only approximate calculations are feasible, and therefore defenders of this theory will always be able to argue that unique predictions might be possible if more refined calculations were carried out in the future.

Fortunately, the hypothesis of formative causation differs from the conventional theory in a second important respect. According to the latter, the causes which give rise to a form on the first occasion, or on the hundredth, or the billionth should operate in exactly the same way, since they are assumed to be changeless. The same expectation follows from theories which seek to account for empirically observable forms in terms of eternal archetypal Forms or transcendent mathematical truths. But according to the hypothesis of formative causation, the form of a system depends on the cumulative morphic influence of previous similar systems. Thus this influence will be stronger on the billionth occasion than on the thousandth or the tenth. If this cumulative aspect of formative causation could be demonstrated empirically, the hypothesis could be distinguished both from the conventional theory and from theories of the Platonic and Pythagorean types.

In the case of morphic units which have existed for a very long time – thousands of millions of years in the case of the hydrogen atom – the morphogenetic field will be so well established as to be effectively changeless. Even the morphogenetic fields of morphic units which originated as recently as a few centuries or decades ago may be subject to the influence of so many past systems that any increments in this influence will be too small to be detectable. But with brand-new forms, it may well be possible to detect the cumulative morphic influence experimentally.

Consider a newly synthesized organic chemical which has never existed before. According to the hypothesis of formative causation, its crystalline form will not be predictable in advance, and no morphogenetic field for this form will yet exist. But after it has been crystallized for the first time, the form of its crystals will influence subsequent crystallizations by morphic resonance, and the oftener it is crystallized, the stronger should this influence become. Thus on the first occasion, the substance may not crystallize at all readily; but on subsequent occasions crystallization should occur more and more easily as increasing numbers of past crystals contribute to its morphogenetic field by morphic resonance.

In fact, chemists who have synthesized entirely new chemicals often have great difficulty in getting these substances to crystallize for the first time. But as time goes on, these substances tend to crystallize with greater and greater ease.

This principle is illustrated in the following account, taken from a textbook on crystals, of the spontaneous and unexpected appearance of a new type of crystal:

> 'About ten years ago a company was operating a factory which grew large single crystals of ethylene diamine tartrate from solution in water. From this plant it shipped the crystals many miles to another which cut and polished them for industrial use. A year after the factory opened, the crystals in the growing tanks began to grow badly; crystals of something else adhered to them – something which grew even more rapidly. The affliction soon spread to the other factory: the cut and polished crystals acquired the malady on their surfaces . . .
>
> The wanted material was *anhydrous* ethylene diamine tartrate, and the unwanted material turned out to be the *monohydrate* of that substance. During three years of research and development, and another year of manufacture, no seed of the monohydrate had formed. After that they seemed to be everywhere.' (A. Holden and P. Singer.[7])

These authors suggest that on other planets, types of crystal which are common on earth may not yet have appeared, and add: 'Perhaps in our own world many other possible solid species are still unknown, not because their ingredients are lacking, but simply because suitable seeds have not yet put in an appearance.'[8]

The conventional explanation of the fact that substances usually crystallize more easily after they have been crystallized for the first time, and of the fact that the ease of crystallization generally increases the more often they are crystallized is that fragments of previous crystals 'infect' subsequent solutions. When there is no obvious means by which these seeds could have moved from place to place, they are assumed to have travelled through the atmosphere as microscopic dust particles.

There can be no doubt that the 'infection' of a supersaturated solution with appropriate crystal seeds greatly facilitates crystallization. But according to the hypothesis of formative causation, the crystallization of a substance should also be facilitated by the mere fact that it has crystallized before. So when substances are found to crystallize more readily the oftener they are crystallized, an increasing number of invisible seeds in the atmosphere may not be the only explanation. This question could be investigated experimentally under conditions in which dust particles were removed by filtering the air and all other potential contamination was eliminated. The period of time taken by a newly synthesized substance to crystallize from a supersaturated solution could be measured under rigorously standardized conditions in the absence of seeds both before and after this substance had been crystallized repeatedly elsewhere. A decrease in this period would provide evidence in favour of the hypothesis of formative causation.

In more complicated experiments, it might be possible to demonstrate not only that the morphogenetic field of a particular crystal species was subject to the cumulative influence of past crystals, but also that the structure of this field was not determined before the first crystal of this type appeared. For instance, consider the following procedure.

A solution of a newly synthesized chemical is divided into several batches, say P, Q and R, each of which is taken to a different laboratory located hundreds of miles away from the other laboratories as an additional precaution to avoid cross-contamination by seeds. Now each batch is deliberately seeded with a different type of crystal in an attempt to encourage different patterns of crystallization of the new chemical, whose crystal form is as yet undetermined *ex hypothesi*. These crystallizations take place as far as possible simultaneously. Assume that P, Q and R each gives a different type of crystal. Samples of these crystals are analysed and their structures determined by X-ray crystallography. Now one is selected at random, say R, and large batches of the chemical are repeatedly crystallized using seeds of R-type crystals. According to the hypothesis of formative causation, these large numbers of R-type crystals should have a stronger morphic influence on all subsequent crystallizations than the small initial samples of P- and Q- type crystals, and hence there should be a higher probability of obtaining R-type than P- or Q-type crystals.

An attempt is now made to repeat the P- and Q-type crystallizations with the same sorts of seeds that were used initially. Crystallization is also carried out in the absence of any seeds whatsoever. If in all these cases, R-type crystals are obtained, the result would strongly support the hypothesis of formative causation. And if this type of experiment could be repeated with many different newly synthesized substances, a really convincing weight of evidence could be built up.

However, if only a single type of crystal were obtained initially in P, Q and R the result would be inconclusive. On the one hand, if crystallization began slightly sooner in one of these solutions than in the others, the influence of these crystals by morphic resonance might be strong enough to cause the same type of crystallization to occur in the other solutions. On the other hand, this result would also agree with the conventional assumption that a single crystal-form would be obtained because it was a unique minimum-energy structure. Nevertheless, even with a single type of crystal, it should still be possible to detect a decrease in the time taken by the substance to crystallize under standard conditions as increasing numbers of past crystals of this type contribute to the morphogenetic field by morphic resonance.

Experiments with crystals are only one of the ways in which the hypothesis of formative causation could be tested. Examples of possible experiments with biological systems are discussed in Sections 7.4, 7.6, 11.2, and 11.4.

Notes

1 Mackie (1974), p.19.

2 Hesse (1961), p.285.

3 Many examples of oscillations within biological systems have been described. See, for example, the review of oscillations at the cellular level by Rapp (1979).

4 The vibration of a system brought about by a 'one-dimensional' energetic stimulus can in fact give rise to definite forms and patterns:

simple examples are the 'Chladni figures' produced by sand or other small particles on a vibrating diaphragm. Illustrations of numerous two- and three-dimensional patterns on vibrating surfaces can be found in Jenny (1967). But this is not comparable to the type of morphogenesis brought about through morphic resonance.

5 For discussions of the possibility of causal influences from future events, see Hesse (1961) and Mackie (1974).

6 Evidence for precognition would be relevant to this argument only if mental states were assumed, on metaphysical grounds, to be an aspect of physical states of the body, or to run parallel to them, or to be epiphenomena of them. However, from the point of view of Interactionism, an influence from future *mental* states would not necessarily require a *physical* influence to pass 'backwards' in time. These metaphysical alternatives are discussed further in Chapter 12.

7 Holden and Singer (1961), pp.80-81.

8 ibid., p.81.

6 Formative Causation and Morphogenesis

6.1 Sequential morphogeneses

After sub-atomic particles have aggregated into atoms, the atoms may combine together into molecules, and the molecules into crystals. The crystals then retain their form indefinitely as long as the temperature remains below their melting point. By contrast, in living organisms morphogenetic processes continue indefinitely in the endlessly repeated cycles of growth and reproduction.

The simplest living organisms consist of single cells, the growth of which is followed by division, and division by growth. Thus the morphogenetic germs for the chreodes of division must appear within the final forms of the fully grown cells; and the newly divided cells serve as the starting points for the chreodes of cellular growth and development.

In multicellular organisms, these cycles continue in only some of the cells, for example in germ cell lines, 'stem' cells, and meristematic cells. Other cells, and indeed whole tissues and organs, develop into a variety of specialized structures which undergo little further morphogenetic change: they stop growing, although they may retain the ability to regenerate after damage; and sooner or later they die. In fact, they may be mortal precisely because they cease to grow.[1]

The development of multicellular organisms takes place through a series of stages controlled by a succession of morphogenetic fields. At first the embryonic tissues develop under the control of primary embryonic fields. Then sooner (in 'mosaic' development) or later (in 'regulative' development), different regions come under the influence of secondary fields, in animals those of limbs, eyes, ears, etc.; in plants of leaves, petals, stamens, etc. Generally speaking, the morphogenesis brought about by the primary fields is not spectacular, but it is of fundamental importance because it establishes the characteristic differences between cells in different regions which (according to the present hypothesis) enable them to act as the morphogenetic germs of the organ fields. Then in the tissues developing under their influence, germs of subsidiary fields appear, fields which control the morphogenesis of structures within the organ as a whole: in a leaf, the lamina, stipules, petiole, etc.; in an eye, the cornea, iris, lens etc. And then still lower-level morphogenetic fields come into play: for example those for vascular differentiation within the lamina of a leaf and for the differentiation of stomata and hair cells on its surface.

These fields can be, and have been, investigated experimentally by studying the ability of developing organisms to regulate after damage to different regions of the embryonic tissue, and after grafting tissue taken from one region into another. Both in animal embryos and in the meristematic zones of plants, as the development of the tissues proceeds the different regions behave with increasing autonomy; the system as a whole loses the ability to regulate, but local regulations occur within the developing organs as the primary embryonic fields are supplanted by more numerous secondary fields.[2]

6.2 The polarity of morphogenetic fields

Most biological morphic units are polarized in at least one direction. Their morphogenetic fields, containing polarized virtual forms, will automatically take up appropriate orientations if their morphogenetic germs are also intrinsically polarized; but if they are not, a polarity must first be imposed on them. For example, the spherical egg cells of the alga *Fucus* have no inherent polarity, and their development can begin only after they have been polarized by any one of a variety of directional stimuli including light, chemical gradients and electric currents; in the absence of any such stimuli, a polarity is taken up at random, presumably owing to chance fluctuations.

Nearly all multicellular organisms are polarized in a shoot-root or head-tail direction, many also in a second direction (ventral-dorsal), and some in all three (head-tail, ventral-dorsal, and left-right). The latter are consequently asymmetrical and potentially capable of existing in forms which are mirror-images of each other; for example snails with spiral shells. And in organisms which are bilaterally symmetrical, asymmetrical structures which are borne on both sides necessarily occur in both right and left 'handed' forms, for example right and left hands.

These mirror-image forms have the same morphology, and they presumably develop under the influence of the same morphogenetic field. The field simply takes on the handedness of the morphogenetic germ with which it becomes associated. Thus both right and left handed previous systems influence both right and left handed subsequent systems by morphic resonance.

This interpretation is supported by some well-known facts of biochemistry. The molecules of amino acids and sugars are asymmetric and are capable of existing in both left and right 'handed' forms. Yet in living organisms, all the amino acids in proteins are left handed, while most of the sugars are right handed. The perpetuation of these chemical asymmetries is made possible by the asymmetric structures of the enzymes which catalyse the synthesis of the molecules. In nature, most of the amino acids and sugars occur rarely, if at all, outside living organisms. Therefore these particular asymmetric forms should contribute overwhelmingly by morphic resonance to the morphogenetic fields of the molecules. But when they are synthesized artificially equal proportions of right and left handed forms are obtained, indicating that the morphogenetic fields have no intrinsic handedness.

6.3 The size of morphogenetic fields

The dimensions of particular atomic and molecular morphic units are more or less constant; so are those of crystal lattices, although they are repeated indefinitely to give crystals of different sizes. Biological morphic units are more variable: not only are there differences between cells, organs, and organisms of given types, but individual morphic units themselves change size as they grow. If morphic resonance is to take place from past systems with similar forms but different sizes, and if a particular morphogenetic field is to remain associated with a growing system, then forms must be capable of being 'scaled up' or 'scaled down' within the morphogenetic field. Thus their essential features must depend not on the absolute but on the relative positions of their component parts, and on their relative rates of vibration. A simple analogy is provided by the music produced by playing a gramophone record at different speeds: it remains recognizable in spite of absolute alterations in all the pitches and rhythms because the relations of the notes and rhythms to each other remain the same.

Although morphogenetic fields may be adjustable in absolute size, the range within which the size of a system can vary is limited by severe physical constraints. In three-dimensional systems, changes in surface area and volume vary respectively as the square and cubic powers of the linear dimensions. This simple fact means that biological systems cannot be magnified or diminished indefinitely without becoming unstable.[3]

6.4 The increasing specificity of morphic resonance during morphogenesis

Energetic resonance is not an 'all or none' process: a system resonates in response to a range of frequencies which are more or less close to its natural frequency, although the maximum response occurs only when the frequency coincides with its own. Analogously, morphic resonance may be more or less finely 'tuned', occurring with greatest specificity when the forms of past and present systems are most closely similar.

When a morphogenetic germ comes into morphic resonance with the forms of countless previous higher-level systems, these forms do not coincide exactly but give rise to a probability structure. As the first stages of morphogenesis take place, structures are actualized at particular places within the regions given by the probability structure. The system now has a more developed and better-defined form, and will consequently resemble the forms of some previous similar systems more closely than others; the morphic resonance from these forms will be more specific and hence more effective. And as development proceeds the selectivity of morphic resonance will increase still further.

A very general illustration of this principle is given by the development of a organism from a fertilized egg. The early stages of embryology often resemble those of numerous other species or even families and orders. As development proceeds, the particular features of the order, family, genus and finally species tend to appear sequentially, and the relatively minor differences that distinguish the individual organism from other individuals of the same species generally appear last.

This increasingly specific morphic resonance will tend to canalize development towards particular variants of the final form which were expressed in previous organisms. The detailed pathway of development will be affected by both genetic and environmental factors: an organism of a particular genetic constitution will tend to develop in such a way that it enters into specific morphic resonance with previous individuals with the same genetic constitution; and environmental effects on development will tend to bring the organism under the specific morphic influence of previous organisms which developed in the same environment.

Previous similar morphic units which were part of the same organism will have an even more specific effect. For example, in the development of leaves on a tree, the forms of previous leaves on the same tree are likely to make a particularly significant contribution to the morphogenetic field, tending to stabilize the leaf-form characteristic of this particular tree.

6.5 The maintenance and stability of forms

At the end of a process of morphogenesis, the actual form of a system comes into coincidence with the virtual form given by the morphogenetic field. The continued association of the system with its field is revealed most clearly in the phenomenon of regeneration. The restoration of the form of the system after small deviations from the final form is less obvious, but no less important: the morphic unit is continuously stabilized by its morphogenetic field. In biological systems, and to some extent in chemical systems, this maintenance of form enables the morphic units to persist even though their constituent parts change as they are 'turned over' and replaced. The morphogenetic field itself persists owing to the continuing influence of the forms of similar past systems.

An extraordinarily interesting feature of the morphic resonance acting on a system with a persisting form is that this resonance will include a contribution from the past states of the system itself. In so far as a system resembles itself in the past more closely than it resembles any other past system, this self-resonance will be highly specific. It may in fact be of the most fundamental importance in maintaining the very identity of the system.

Matter can no longer be thought of as made up of solid particles like tiny billiard balls that endure throughout time. Material systems are dynamic structures which are constantly re-creating themselves. On the present hypothesis, the persistence of material forms depends on a continuously repeated actualization of the system under the influence of its morphogenetic field; at the same time the morphogenetic field is continuously re-created by morphic resonance from similar past forms. The forms which are most similar and which will consequently have the greatest effect will be those of the system itself in the immediate past. This conclusion would appear to have profound physical implications: the preferential resonance of a system with itself in the immediate past could conceivably help to account for its persistence not only in time, but also at a particular place.[4]

6.6 A note on physical 'dualism'

All actual morphic units can be regarded as *forms of energy*. On the one hand, their structures and patterns of activity depend on the morphogenetic fields with which they are associated, and under the influence of which they have come into being. On the other hand, their very existence and their ability to interact with other material systems is due to the energy bound within them. But although these aspects of form and energy can be separated conceptually, in reality they are always associated with each other. No morphic unit can have energy without form, and no material form can exist without energy.

This physical 'duality' of form and energy which is made explicit by the hypothesis of formative causation has much in common with the so-called wave-particle duality of quantum theory.

According to the hypothesis of formative causation, there is only a difference of degree between the morphogenesis of atoms and that of molecules, crystals, cells, tissues, organs and organisms. If 'dualism' is defined in such a way that the orbitals of electrons in atoms involve a duality of waves and particles, or of form and energy, then so do the more complex forms of higher-level morphic units; but if the former are not considered to be dualistic, then neither are the latter.[5]

In spite of their similarity, there is of course a difference in kind between the hypothesis of formative causation and the conventional theory. The latter provides no fundamental understanding of the causation of forms, unless equations or 'mathematical structures' describing them are assumed to play a causal role; if so, a very mysterious dualism between mathematics and reality would appear to be implied. The hypothesis of formative causation overcomes this problem by regarding the forms of previous systems as the causes of subsequent similar forms. From the conventional point of view, this cure may seem worse than the disease in so far as it requires an action across time and space unlike any known type of physical action. However, this is not a metaphysical but a physical proposition, and is capable of being tested experimentally.

If this hypothesis is supported by experimental evidence, then not only might it allow the various matter fields of quantum field theory to be interpreted in terms of morphogenetic fields, but it could also lead towards a new understanding of other physical fields.

In the morphogenetic field of an atom, a naked atomic nucleus surrounded by virtual orbitals serves as a morphogenetic 'attractor' for electrons. Perhaps the so-called electrical attraction between the nucleus and the electrons could be regarded as an aspect of this atomic morphogenetic field. When the final form of the atom has been actualized by the capture of electrons, it no longer acts as a morphogenetic 'attractor', and in electrical terminology it is neutral. So it is not inconceivable that electromagnetic fields could be derived from the morphogenetic fields of atoms.

In a comparable manner, it might eventually be possible to interpret the strong and weak nuclear forces in terms of the morphogenetic fields of atomic nuclei and nuclear particles.

6.7 A summary of the hypothesis of formative causation

(i) In addition to the types of energetic causation known to physics, and in addition to the causation due to the structures of known physical fields, a further type of causation is responsible for the forms of all material morphic units (sub-atomic particles, atoms, molecules, crystals, quasi-crystalline aggregates, organelles, cells, tissues, organs, organisms). Form, in the sense used here, includes not only the shape of the outer surface of the morphic unit but also its internal structure. This causation, called *formative causation*, imposes a spatial order on changes brought about by energetic causation. It is not itself energetic, nor is it reducible to the causation brought about by known physical fields. (Sections 3.3, 3.4.)

(ii) Formative causation depends on *morphogenetic fields*, structures with morphogenetic effects on material systems. Each kind of morphic unit has its own characteristic morphogenetic field. In the morphogenesis of a particular morphic unit, one or more of its characteristic parts – referred to as the *morphogenetic germ* – becomes surrounded by, or embedded within, the morphogenetic field of the entire morphic unit. This field contains the morphic unit's virtual form, which is actualized as appropriate component parts come within its range of influence and fit into their appropriate relative positions. The fitting into position of the parts of a morphic unit is accompanied by a release of energy, usually as heat, and is thermodynamically spontaneous; from an energetic point of view, the structures of morphic units appear as minima or 'wells' of potential energy. (Sections 3.4, 4.1, 4.2, 4.4, 4.5.)

(iii) Most inorganic morphogenesis is rapid, but biological morphogenesis is relatively slow and passes through a succession of intermediate stages. A given type of morphogenesis usually follows a particular developmental pathway; such a canalized pathway of change is called a *chreode*. Nevertheless, morphogenesis may also proceed towards the final form from different morphogenetic germs and by different pathways, as in the phenomena of regulation and regeneration. In the cycles of cell growth and cell division and in the development of the differentiated structures of multicellular organisms, a succession of morphogenetic processes take place under the influence of a succession of morphogenetic fields. (Sections 2.4, 4.1, 5.4, 6.1.)

(iv) The characteristic form of a given morphic unit is determined by the forms of previous similar systems which act upon it across time and space by a process called *morphic resonance*. This influence takes place through the morphogenetic field and depends on the systems' three-dimensional structures and patterns of vibration. Morphic resonance is analogous to energetic resonance in its specificity, but it is not explicable in terms of any known type of resonance, nor does it involve a transmission of energy. (Sections 5.1, 5.3.)

(v) All similar past systems act upon a subsequent similar system by morphic resonance. This action is provisionally assumed not to be attenuated by space or time, and to continue indefinitely; however the relative effect of a given system declines as the number of similar systems contributing to morphic resonance increases. (Sections 5.4, 5.5.)

(vi) The hypothesis of formative causation accounts for the repetition of forms but does not explain how the first example of any given form originally came into being. This unique event can be ascribed to chance, or to a creativity inherent in matter, or to a transcendent creative agency. A decision between these alternatives can be made only on metaphysical grounds and lies outside the scope of the hypothesis. (Section 5.1.)

(vii) Morphic resonance from the intermediate stages of previous similar processes of morphogenesis tends to canalize subsequent similar morphogenetic processes into the same chreodes. (Section 5.4.)

(viii) Morphic resonance from past systems with a characteristic polarity can only occur effectively after the morphogenetic germ of a subsequent system has been suitably polarized. Systems which are asymmetrical in all three dimensions and exist in right or left 'handed' forms influence subsequent similar systems by morphic resonance irrespective of handedness. (Section 6.2.)

(ix) Morphogenetic fields are adjustable in absolute size and can be 'scaled up' or 'scaled down' within limits. Thus previous systems influence subsequent systems of similar form by morphic resonance even though their absolute sizes may differ. (Section 6.3.)

(x) Even after adjustment for size, the many previous systems influencing a subsequent system by morphic resonance are not identical, but only similar in form. Therefore their forms are not precisely superimposed within the morphogenetic field. The most frequent type of previous form makes the greatest contribution by morphic resonance, the least frequent the least: morphogenetic fields are not precisely defined but are represented by *probability structures* which depend on the statistical distribution of previous similar forms. The probability distributions of electronic orbitals described by solutions of the Schrödinger equation are examples of such probability structures, and are similar in kind to the probability structures of the morphogenetic fields of morphic units at higher levels. (Sections 4.3, 5.4.)

(xi) The morphogenetic fields of morphic units influence morphogenesis by acting upon the morphogenetic fields of their constituent parts. Thus the fields of tissues influence those of cells; those of cells, organelles; those of crystals, molecules; those of molecules, atoms; and so on. These actions depend on the influence of higher-level probability structures on lower-level probability structures and are thus inherently probabilistic. (Sections 4.3, 4.4.)

(xii) Once the final form of a morphic unit is actualized, the continued action of morphic resonance from similar past forms stabilizes and maintains it. If the form persists, the morphic resonance acting upon it will include a contribution from its own past states. In so far as the system resembles its own past states more closely than those of other systems, this morphic resonance will be highly specific, and may be of considerable importance in maintaining the system's identity. (Sections 6.4, 6.5.)

(xiii) The hypothesis of formative causation is capable of being tested experimentally. (Section 5.6.)

Notes

1 It seems probable that an important cause of ageing, at least at the cellular level, is the build-up of harmful waste products which cells are

unable to eliminate. According to a recent theory, if cells grow fast enough they can keep 'one step ahead' of this build-up simply because these substances are diluted by growth. Furthermore, in asymmetric cell divisions, which are common in higher animals and plants, these substances may be passed on unequally to the daughter cells: one may be rejuvenated at the expense of the increased mortality of the other. Thus rejuvenation depends on growth and cell division: morphogenetic end-points – the differentiated cells, tissues and organs of multicellular organisms – are necessarily mortal (Sheldrake, 1974).

2 For animal examples, see Weiss (1939); for plants, Wardlaw (1965).

3 The classical discussion of this elementary but important point is in the chapter 'On Magnitude' in Thompson (1942).

4 If the system 'identifies' itself with a particular location and if its persistence at this location depends on morphic resonance with itself in the immediate past, its resistance to being moved from this location – its *inertial mass* – should be related to the frequency with which this self-resonance occurs. For resonance depends on characteristic cycles of vibration; it cannot occur in an instant, because a cycle of vibration takes time. The higher the frequency of vibration, the more recent will be the past states with which self-resonance occurs; thus the greater will be the tendency of the system to be 'tied' to its position in the immediate past. Conversely, the lower the frequency of vibration the less strong will be the tendency of a system to 'identify' itself with itself at a particular location: it will be able to move further relative to other objects before it 'notices' that it has done so.

There is a remarkable resemblance between the relationship suggested above and the proportionality between the mass of a particle and the frequency of its matter wave given by the de Broglie equation:

$$m = \frac{hv}{c^2}$$

where m is the mass of the particle, v the frequency of vibration, h Planck's constant, and c the velocity of light. This relationship is fundamental to quantum mechanics and is amply supported by experimental evidence.

5 Sir Karl Popper, among others, has argued that talking of a dualism of particle and wave has led to much confusion, and has suggested that the term dualism should be abandoned:

> 'I propose that we speak instead (as did Einstein) of the particle and its *associated* propensity fields (the plural indicates that the fields depend not only on the particle but also on other conditions), thus avoiding the suggestion of a symmetrical

> relation. Without establishing some such terminology as this ('association' in place of 'dualism'), the term 'dualism' is bound to survive, with all the misconceptions connected with it; for it does point to something important: the association that exists between particles and fields of propensities.' (Popper, 1967, p.41.)

This proposal would appear to harmonize well with the hypothesis of formative causation if propensity fields are taken to include morphogenetic fields.

7 The Inheritance of Form

7.1 Genetics and heredity

Hereditary differences between otherwise similar organisms depend on genetic differences; genetic differences depend on differences in the structure of DNA, or in its arrangement within the chromosomes; and these differences lead to changes in the structure of proteins, or to changes in the control of protein synthesis.

These fundamental discoveries, supported by a large body of detailed evidence, provide a satisfyingly straightforward understanding of the inheritance of proteins and of characteristics that depend more or less directly on particular proteins, for example sickle-cell anaemia and hereditary defects of metabolism. By contrast, hereditary differences of form generally bear no immediate and obvious relationship to changes in the structure or synthesis of particular proteins. Nevertheless, such changes could affect morphogenesis in various ways through effects on metabolic enzymes, hormone-synthesizing enzymes, structural proteins, proteins in cell membranes, and so on. Many examples of these effects are already known. But granted that various chemical changes lead to alterations or distortions of the normal pattern of morphogenesis, what determines the normal pattern of morphogenesis itself?

According to the mechanistic theory, cells, tissues, organs and organisms take up their appropriate forms as a result of the synthesis of the right chemicals in the right places at the right times. Morphogenesis is supposed to proceed spontaneously as a result of complex physico-chemical interactions in accordance with the laws of physics. But which laws of physics? The mechanistic theory simply leaves the question open (Section 2.2).

The hypothesis of formative causation suggests a new way of

answering this question. In so far as it offers an interpretation of biological morphogenesis which stresses the analogy with physical processes such as crystallization, as well as ascribing an important role to energetically indeterminate fluctuations, it fulfills rather than denies the expectations of the mechanistic theory. But whereas the latter attributes practically all the phenomena of heredity to the genetical inheritance embodied in the DNA, according to the hypothesis of formative causation organisms also inherit the morphogenetic fields of past organisms of the same species. This second type of inheritance takes place by morphic resonance and not through the genes. So heredity includes *both* genetic inheritance *and* morphic resonance from similar past forms.

Consider the following analogy. The music which comes out of the loudspeaker of a radio set depends *both* on the material structures of the set and the energy which powers it *and* on the transmission to which the set is tuned. The music can of course be affected by changes in the wiring, transistors, condensers, etc., and it ceases when the battery is removed. Someone who knew nothing about the transmission of invisible, intangible and inaudible vibrations through the electromagnetic field might therefore conclude that it could be explained entirely in terms of the components of the radio, the way in which they were arranged, and the energy on which their functioning depended. If he ever considered the possibility that anything entered from outside, he would dismiss it when he discovered that the set weighed the same switched on and switched off. He would therefore have to suppose that the rhythmic and harmonic patterns of the music arose within the set as a result of immensely complicated interactions among its parts. After careful study and analysis of the set, he might even be able to make a replica of it which produced exactly the same sounds as the original, and would probably regard this result as a striking proof of his theory. But in spite of his achievement, he would remain completely unaware that in reality the music originated in a broadcasting studio hundreds of miles away.

In terms of the hypothesis of formative causation, the 'transmission' would come from previous similar systems, and its 'reception' would depend on the detailed structure and organization of the receiving system. As in a radio set, two types of change in the organization of the 'receiver' would have significant effects. First, changes in the 'tuning' of the system could lead to the reception of quite different 'transmissions': just as a radio set can be tuned to different radio stations, so a developing system can be 'tuned' to different morphogenetic fields. Second, just as changes within a radio set tuned to a particular station can lead to modifications and distortions of the music coming out of the loudspeaker, so changes within a system developing under the influence of a particular morphogenetic field can lead to various modifications and distortions of the final form.

Thus in developing organisms both environmental and genetic factors could affect morphogenesis in two different ways: either by changing the 'tuning' of morphogenetic germs, or by changing the usual pathways of morphogenesis in such a way that variants of the normal final forms are produced.

7.2 Altered morphogenetic germs

The morphogenetic germs for the development of organs and of tissues consist of cells or groups of cells with characteristic structures and patterns of oscillation (Sections 4.5, 6.1). If as a result of unusual environmental conditions or genetic alterations the structure and oscillatory pattern of a germ were changed sufficiently, it would no longer become associated with the usual higher-level field: either it might fail to act as a germ at all, in which case an entire structure would fail to appear within the organism; or it might become associated with a different morphogenetic field, in which case a structure not normally found in this part of the organism would develop instead of the usual one.

Many examples of such a loss of an entire structure or of the replacement of one structure by another have been described. Sometimes the same changes can be brought about by genetic factors and by changes in the environment of the developing organism; the latter are referred to in the genetical literature as 'phenocopies'.

Effects of these types have been studied in great detail in the fruit fly *Drosophila*. A considerable number of identified mutations lead to transformations of entire regions of the fly; for example 'antennapedia' changes the antennae into legs, and mutations within the 'bithorax' gene complex cause the metathoracic segment, which normally bears two halteres, to develop as if it were a mesothoracic segment (Fig. 17). The resulting organisms bear two pairs of wings on adjacent segments.[1]

Comparable phenomena have been found in plants. In the pea, for example, the leaves normally bear leaflets towards their base and tendrils at their tip. In some leaves tendrils are formed opposite leaflets, indicating that similar primordia are capable of giving rise to both types of structure (Fig. 18); presumably cells within these

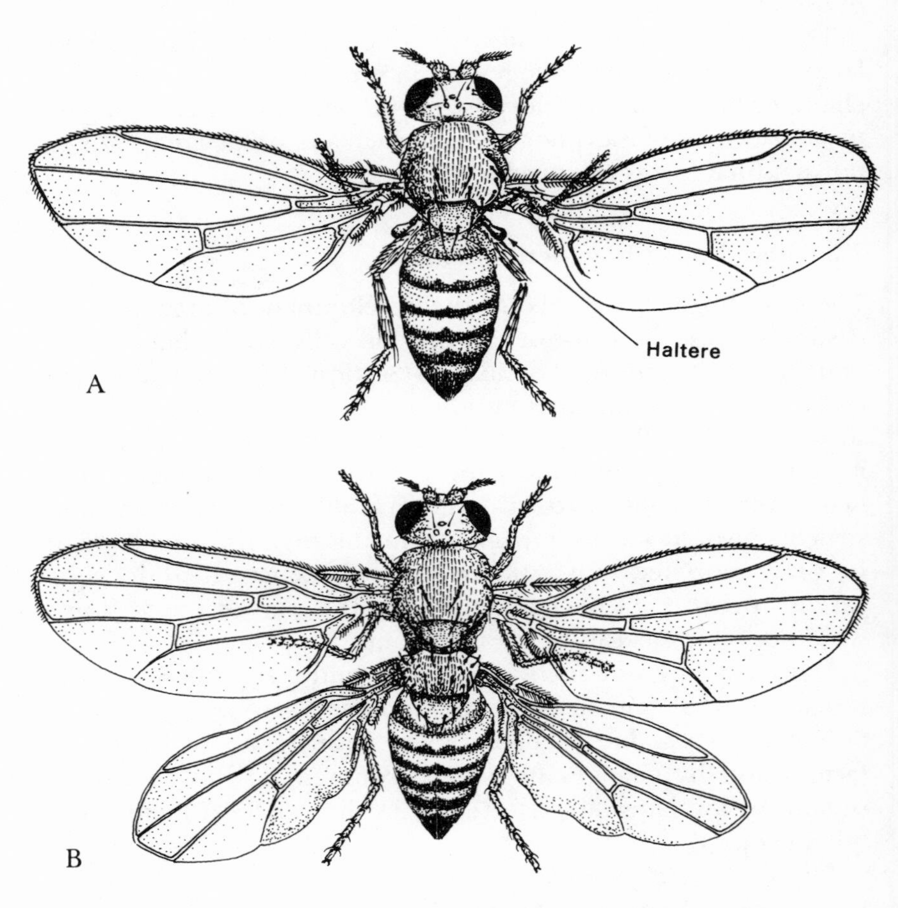

Figure 17 A normal specimen of the fruit fly *Drosophila* (A) and a mutant fly (B) in which the third thoracic segment has been transformed in such a way that it resembles the second thoracic segment. The fly consequently has two pairs of wings instead of one.

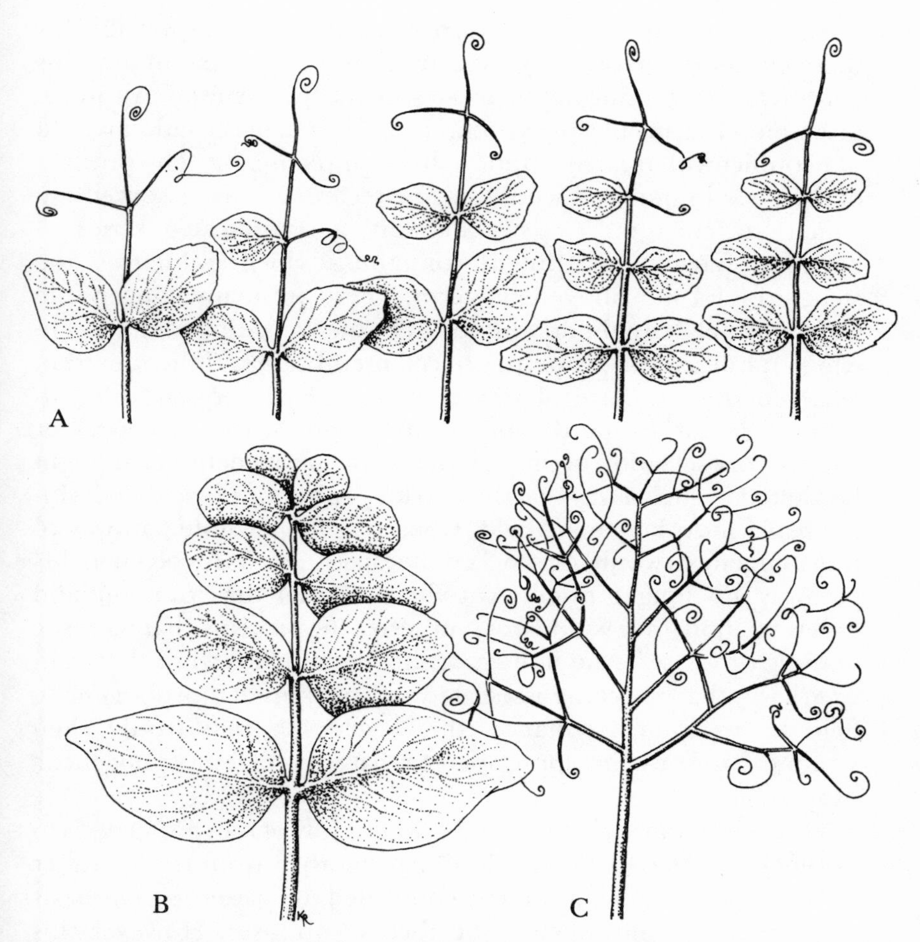

Figure 18 A: Normal pea leaves, bearing both leaflets and tendrils.
B: Leaf of a mutant plant in which only leaflets are formed.
C: Leaf of a mutant plant in which only tendrils are formed.

primordia are influenced by factors within the embryonic leaf causing them to take up the structure and oscillatory pattern characteristic of the morphogenetic germ either of a tendril, or of a leaflet. However, in one type of mutant, the ability to form tendrils is suppressed and all the primordia give rise to leaflets; in another mutant (due to a gene on a different chromosome) the formation of leaflets is suppressed and all the primordia give rise to tendrils[2] (Fig. 18).

The conventional interpretation is that the genes responsible for these effects are involved in the control of the synthesis of proteins necessary for the normal processes of morphogenesis. An interpretation in terms of the hypothesis of formative causation would not conflict with this assumption, but complement it. The product of the gene in question would not be regarded only as something which 'switched on' or 'switched off' a complicated series of chemical interactions, but as something which influenced the structure of a morphogenetic germ. There are many conceivable ways in which it might do this, for example by coding for a protein which modified the properties of cell membranes. If the mutation changed the structure of this protein and consequently led to changes in the properties of the membranes, the structures or patterns of oscillation of the cells of the morphogenetic germ might be altered in such a way that they no longer became associated with the usual morphogenetic field. Consequently a whole pathway of morphogenesis would be blocked. Because the cells involved in this pathway no longer underwent their normal development and differentiation, they would not synthesize the proteins characteristic of these processes. And if the morphogenetic germ were altered in such a way that it became associated with a different morphogenetic field by morphic resonance, the developing cells would then synthesize the proteins appropriate to that particular morphogenetic process.

Thus a mutation which caused one pathway of morphogenesis to be blocked, or led to the substitution of another, would indeed alter a gene product which indirectly controlled the patterned synthesis of proteins, as the mechanistic theory supposes. However this control would not depend on complicated chemical interactions alone, but would be mediated by morphogenetic fields.

7.3 Altered pathways of morphogenesis

Whereas the factors affecting morphogenetic germs have *qualitative* effects on morphogenesis resulting in the absence of a structure, or the substitution of one structure for another, many environmental and genetic factors bring about *quantitative* modifications of the final forms of structures through their effects on the processes of morphogenesis. For example, in plants of a given cultivated variety grown under a range of environmental conditions, the overall shape of the shoot and root systems, the morphology of the leaves, and even the anatomy of various organs differ in detail; but nevertheless the characteristic varietal form remains recognizable. Or in different varieties of the same species grown in the same environment, the plants differ from each other in many details, although they are all recognizably variants of a characteristic specific form: species are, after all, primarily defined in terms of their morphology.

Genetic and environmental factors influence development through various quantitative effects on structural components, enzymic activity, hormones, etc. (Section 7.1.) Some of these influences are relatively unspecific and affect several different pathways of morphogenesis. Others may perturb the normal course of development but have little effect on the final form, owing to regulation.

While certain striking genetic effects may be traceable to particular genes, most depend on the combined influence of numerous genes, the individual effects of which are small, and difficult to identify and analyse.

According to the hypothesis of formative causation, organisms of the same variety or race will resemble each other not only because they are genetically similar and therefore subject to similar genetic influences during morphogenesis, but also because their characteristic varietal chreodes are reinforced and stabilized by morphic resonance from past organisms of the same variety.

The morphogenetic fields of a species are not fixed, but change as the species evolves. The greatest statistical contribution to the probability structures of the morphogenetic fields will be made by the most common morphological types, which will also be those which developed under the most usual environmental conditions. In the simplest cases, the automatic averaging effect of morphic resonance will stabilize the morphogenetic fields around a single most probable form or 'wild type'. But if the species inhabits two or more geographically or ecologically distinct environments in which characteristic varieties or races have evolved, the morphogenetic fields of the species will not contain a single most probable form, but a 'multi-modal' distribution of forms, depending on the number of morphologically distinct varieties or races and the relative sizes of their past populations.

7.4 Dominance

At first sight, the idea that varietal forms are stabilized by morphic resonance from past organisms of the same variety may appear to add little to the conventional explanation in terms of genetic similarity alone. However, its importance becomes apparent in considering hybrid organisms which are subject to morphic resonance from two distinct parental types.

To return to the radio analogy: under normal circumstances a set is tuned to only one station at a time, just as an organism is normally 'tuned' to similar past organisms of the same variety. But if the radio is tuned into two different stations simultaneously, the sounds it produces depend on the relative strength of their signals: if one is very strong and the other very weak, the latter has little noticeable effect; but if both are of similar strength, the set produces a mixture of sounds from both sources. Likewise, in a hybrid produced by crossing two varieties, the presence of genes and gene-products characteristic of both will tend to bring the developing organism into morphic resonance with past organisms of both parental types. Now the overall probability structures within the morphogenetic fields of the hybrid will depend on the relative strength of the morphic resonance from the two parental types. If both parents come from varieties represented by comparable numbers of past individuals, both will tend to influence morphogenesis to similar extents, giving a combination or 'resultant' of the two parental forms (Fig. 19A). But if one variety has been represented by fewer individuals than the other, their smaller contribution to the overall probability structure will mean that the form of the other parental variety will tend to predominate (Fig. 19B). And if one of the parents comes from a mutant line of recent origin, morphic resonance from the small number of past individuals of this type will make an insignificant contribution to the probability structure of the hybrids (Fig. 19C).

These expectations are in harmony with the facts. First, hybrids between well-established varieties or species usually combine features of both, or are of intermediate form. Second, in hybrids between a relatively recent variety and a long-established variety, the characters of the latter are usually more or less dominant. And third, recent mutations affecting morphological characters are nearly always recessive.

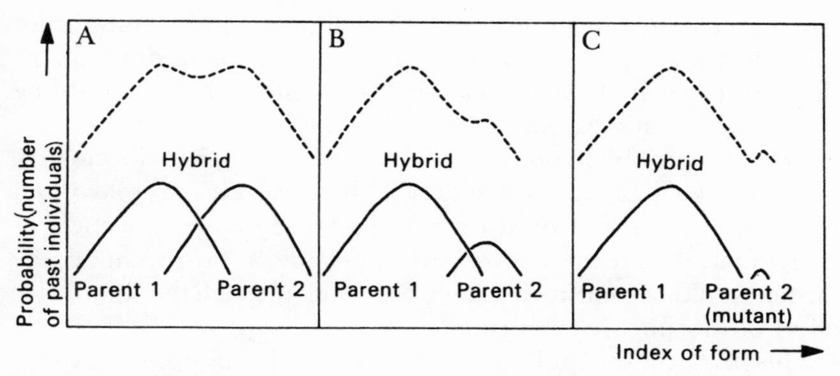

Figure 19 Diagrammatic representation of the probability structures of the morphogenetic fields of parents and hybrids.

Significantly, mechanistic theories of dominance are both vague and speculative, except in the case of characteristics which depend more or less directly on particular proteins. If a mutant gene leads to a loss of function, for example by giving rise to a defective enzyme, it will be recessive, because in hybrids the presence of a normal gene enables the normal enzyme to be produced and hence the normal biochemical reactions occur. However, in some cases the defective gene product might be positively harmful, for example by interfering with the permeability of membranes, in which case the mutation would tend to be both dominant and lethal.

These explanations are satisfactory as far as they go; but in the absence of any mechanistic understanding of morphogenesis, the attempt to account for dominance in the inheritance of form by extrapolation from the molecular level inevitably begs the question.

The conventional genetical theories of dominance are more sophisticated than the purely biochemical theory; they emphasize that dominance is not fixed, but that it evolves. In order to account for the relative uniformity of wild populations, in which most non-lethal mutations are recessive, they assume that the dominance of the 'wild type' has been favoured by natural selection. One theory postulates the selection of genes which modify the dominance of other genes,[3] and another theory the selection of increasingly effective versions of the genes that control the dominant characters in question.[4] But apart from the fact that there is little evidence in favour of either, and some against both, these theories suffer from the defect that they pre-suppose rather than explain dominance: they only provide hypothetical mechanisms by which it could be maintained or increased.[5]

According to the hypothesis of formative causation, dominance would evolve for a fundamentally different reason. Types favoured by natural selection would be represented by larger numbers of individuals than types of lower survival-value; as time went on, the former would become increasingly dominant through the cumulative effect of morphic resonance.

This hypothesis could in principle be distinguished experimentally from all mechanistic theories of dominance. According to the latter, under a given set of environmental conditions dominance depends only on the genetic constitution of a hybrid, whereas according to the former it depends *both* on the genetic constitution *and* on morphic resonance from the parental types. Therefore if the relative strength of the resonance from the parental types changed, the dominance of one over the other would change even if the genetic constitution of the hybrid remained the same.

Consider the following experiment. Hybrid seeds are obtained from a cross between plants of a well-defined variety (P_1) and a mutant line (P_2). Some of these hybrid seeds are placed in cold storage, while others are grown under controlled conditions. The characteristics of the hybrid plants are carefully observed, and the plants themselves are preserved. In these plants the P_1 morphology is completely dominant (Fig. 19C). Then very large numbers of plants of the mutant type (P_2) are grown in the field. Subsequently, some of the hybrids are again grown under the same conditions as they were before, from the same batch of seeds. Because P_2 now makes a greater contribution by morphic resonance, P_1 may be only partially dominant (Fig. 19B). After growing many more P_2 plants, the form of the hybrids may be intermediate between the two parental types (Fig. 19A). Then still more plants of the P_2 type are grown in large numbers; subsequently the hybrids are again grown under the same conditions as the previous hybrids from the same batch of seeds. Now the P_2 type will make a greater contribution by morphic resonance than P_1, and the P_2 morphology will be dominant.

This result would strongly support the morphic resonance

hypothesis of dominance, and would be completely incomprehensible from the point of view of orthodox genetical theory. The only problem with this experiment is that it might be difficult to perform in practice, since if P_1 is a well-established variety that has existed for a very long time – in the case of a wild variety perhaps for many thousands or even millions of years – it would not be practicable to grow comparable numbers of the P_2 type. The experiment would be feasible only if P_1 were a recent variety of which only a relatively small number of individuals had been grown in the past.

7.5 Family resemblances

Within a given variety, organisms differ from each other in all sorts of minor ways. In an interbreeding population each individual is more or less unique genetically, and thus tends to follow its own characteristic path of development under the various quantitative influences of its genes. Moreover, since morphogenesis depends on the effect of probability structures on probabilistic events, the whole process is somewhat indeterminate. And then local environments vary. As a result of all these factors, each individual has a characteristic form and makes its own unique contribution to subsequent morphogenetic fields.

The most specific morphic resonance acting on a particular organism is likely to be that from previous closely-related individuals with a similar genetical constitution, accounting for family resemblances. This specific morphic resonance will be superimposed on the less specific resonance from numerous past individuals of the same variety; and this in turn will be superimposed on a general background of morphic resonance from all past members of the species.

In the valley model of a chreode (cf. Fig. 5), the most specific morphic resonances would determine the detailed course of morphogenesis, corresponding to the bed of a stream, and the less specific morphic resonance from previous individuals of the same variety the bed of a small valley. The variant chreodes of different varieties within the same species would correspond to small divergent or parallel valleys within a larger valley representing the chreode of the species as a whole.

7.6 Environmental effects and morphic resonance

The forms of organisms are influenced to varying degrees by the environmental conditions under which they develop. According to the hypothesis of formative causation, they are also influenced by the environmental conditions under which previous similar organisms developed, because the forms of these organisms contribute to their morphogenetic fields by morphic resonance. In terms of the radio analogy, the music coming out of the loudspeaker is affected not only by changes within the receiver, but also by changes within the broadcasting studio: if an orchestra starts playing a different piece of music, the radio set produces different sounds even though its tuning and internal structures remain the same.

Consider, for example, a new variety of a cultivated species. If very large numbers of plants of this variety are grown in one environment and very few in others, the former will make a much larger contribution to the probability structures of the varietal morphogenetic fields; their form will be the most probable form of the variety and will therefore tend to influence the morphogenesis of all subsequent plants of the same variety, even when they are grown in different environments.

In order to test this prediction, it would perhaps be best to use a new variety of a self-pollinated crop; the plants would be very similar to each other genetically, and there would be no danger of their outcrossing with other varieties. To start with, a few plants would be grown in two very different environments, X and Y, and their morphological characters carefully recorded. Some of the original batch of seeds would be placed in cold storage. Then very large numbers of plants would be grown in environment Y (either in one season, or over several generations). Subsequently, using some of the original seeds which had been kept in cold storage, a few plants would once again be grown in environment X. Their morphogenesis should now be influenced by morphic resonance from the large numbers of genetically similar plants grown in environment Y. Consequently they should show more resemblance to the Y-type morphology than did the original X-type plants. (Of course, for a valid comparison of plants grown in X on different occasions, it would be necessary to ensure that the conditions were practically identical; this would be impossible in the field, but could be achieved relatively easily with an artificially controlled environment in a phytotron.)

If this result were actually obtained, it would provide positive evidence for the hypothesis of formative causation, and would be inexplicable in terms of the mechanistic theory.

A negative result would be inconclusive for two reasons: first, the direct effects of environment X on morphogenetic processes might be so strong that they always chanelled morphogenesis into X-type chreodes in spite of the relatively small stabilizing effect of morphic resonance on these pathways. And second, plants of other varieties of the same species would influence development by morphic resonance, although less specifically; nevertheless this influence might tend to stabilize either or both the X- and the Y-type chreodes, especially if these environments resembled those in which previous varieties of the species normally grew. This effect could be minimized by an appropriate choice of environments.

7.7 The inheritance of acquired characteristics

The influence of previous organisms on subsequent similar organisms by morphic resonance would give rise to effects which could not conceivably occur if heredity depended only on the transfer of genes and other material structures from parents to their progeny. This possibility enables the question of the 'inheritance of acquired characteristics' to be seen in a new light.

In the fierce controversy at the end of the nineteenth and in the earlier part of this century, both the Lamarckians on the one hand, and the followers of Weismann and of Mendel on the other, assumed that heredity depended only on the germ plasm in general or the genes in particular. Therefore if characteristics acquired by organisms in response to the environment were to be inherited, the germ plasm or the genes would have to undergo specific modifications. The anti-Lamarckians emphasized that such modifications seemed extremely unlikely, if not impossible. Even the Lamarckians themselves were unable to suggest any plausible mechanisms by which these changes could be brought about.

On the other hand, the Lamarckian theory seemed to provide a plausible explanation for hereditary adaptations in animals and plants. For example, camels have callosities on their knees. It is easy to understand how these are acquired in response to abrasion of the skin as the camels kneel down. But baby camels are born with them. Facts of this type would make good sense if acquired characteristics somehow became hereditary.

However, the Mendelians deny this possibility, and offer an alternative interpretation in terms of random mutations: if organisms with the acquired characteristics in question are favoured by natural selection, random mutations which happen to produce the same characteristics without the need to acquire them will also be favoured by natural selection, and thus the characteristics will become hereditary. This hypothetical simulation of the inheritance of acquired characteristics is sometimes called the Baldwin effect, after one of the evolutionary theorists who first suggested it.[6]

In the early part of the present century, dozens of scientists claimed to have demonstrated an inheritance of acquired characteristics in various species of animals and plants.[7] The anti-Lamarckians replied with counter-examples, citing again and again the well-known experiment of Weismann, in which he chopped the tails off mice for twenty-two successive generations and found that their progeny still developed tails. Another argument drew attention to the fact that after many generations of circumcision, Jews are still born with foreskins.

After the suicide of one of the leading Lamarckians, P. Kammerer, in 1926, Mendelism became established in the West as the almost unchallenged orthodoxy.[8] Meanwhile, in the Soviet Union believers in the inheritance of acquired characteristics, led by T.D. Lysenko, gained control of the biological establishment in the 1930s and remained dominant until 1964. During this period many of their Mendelian opponents were cruelly persecuted.[9] This polarization resulted in bitterness and dogmatism on both sides.

However, there is now good evidence that acquired characteristics can indeed be inherited; the problem has become one of interpretation. In an important series of experiments, C.H. Waddington exposed the eggs or pupae of wild-type populations of fruit flies to fumes of ether or to high temperatures, causing some of the flies to develop abnormally.[10] The next generation was bred from these abnormal flies, and the eggs or pupae were again exposed to the environmental stress; again the abnormal flies were selected and bred from; and so on. In successive generations, the proportion of abnormal flies increased. After a number of generations (in some cases 14, in others 20 or more) when the progeny of the abnormal flies were raised *without* the environmental stress in a normal environment, some of them still developed the characteristic abnormalities. Moreover these abnormalities continued to appear in their descendants raised under normal conditions. In Waddington's words: 'All these experiments demonstrate that if selection takes place for the occurrence of a character acquired in a particular abnormal environment, the resulting selected strains are liable to exhibit that character even when transferred back to the normal environment.'[11]

Waddington considered the possibility that some physical or chemical influence from the altered structures in the abnormal flies could have induced heritable modifications in their genes,[12] but rejected it because the findings of molecular biology made any such mechanism seem exceedingly unlikely.[13] His final interpretation emphasized both the role of selection for the genetic potential to respond to the environmental stress by developing abnormally, and also the 'canalization of development' involved in the modified morphogenesis. 'To use somewhat picturesque language, one might say that the selection did not merely lower a threshold, but determined in what direction the developing system would proceed once it got over the threshold.'[14] Waddington himself coined the word chreode to express the notion of directed, canalized development. He thought of the determination of the direction taken by a chreode in terms of its 'tuning'. But he did not explain how this canalization and 'tuning' came about, apart from making the vague suggestion that they somehow depended on the selection of genes.[15]

The hypothesis of formative causation complements Waddington's interpretation: the chreodes and the final forms towards which they are directed depend on morphic resonance from previous similar organisms; the 'inheritance of acquired characteristics' of the kind studied by Waddington depends *both* on genetic selection *and* on a direct influence by morphic resonance from the organisms whose development was modified in response to abnormal environments.

In general, pathways of morphogenesis altered by either environmental or genetic factors will tend to canalize and stabilize similar processes of morphogenesis in subsequent similar organisms by morphic resonance. The strength of this influence will depend on the specificity of the resonance and on the number of previous similar organisms whose morphogenesis has been altered; this number will tend to be large if the alterations are favoured by natural or artificial selection, and small if they are not.

Mutilations of fully-formed structures would not alter their pathways of morphogenesis unless they regenerated. Hence mutilations of non-regenerating structures would not be expected to influence the development of subsequent organisms. This conclusion is in agreement with the findings that the amputation of the tails of mice and the circumcision of Jews have no significant hereditary effects.

Notes

1 Morata and Lawrence (1977).
2 Snoad (1974).
3 Fisher (1930).
4 Haldane (1939).
5 Serra (1966).
6 Baldwin (1902).
7 Much of this evidence is summarized by Semon (1912) and Kammerer (1924).
8 Koestler (1971).
9 Medvedev (1969).
10 Papers describing these experiments have been conveniently collected together in Waddington (1975).
11 ibid., p.65.
12 Waddington (1957).
13 See the discussion between C.H. Waddington and A. Koestler in Koestler and Smythies (eds) (1969), pp. 382-391.
14 Waddington (1975), p.87.
15 ibid., pp. 87-88.

8 The Evolution of Biological Forms

8.1 The neo-Darwinian theory of evolution

Very little is actually known, or ever can be known, about the details of evolution in the past. Nor is evolution readily observable in the present. Even on a time scale measured in millions of years, the origin of new species is rare, and of genera, families and orders rarer still. The evolutionary changes which have actually been observed over the last century or so for the most part concern the development of new varieties or races within established species. The best documented examples are of the emergence of dark-coloured races of several European moths in areas where industrial pollution led to the blackening of the surfaces on which they settled. Dark mutants were favoured by natural selection because they were better camouflaged and hence less subject to predation by birds.

With such scanty direct evidence, and with so little possibility of experimental test, any interpretation of the mechanism of evolution is bound to be speculative: unconstrained by detailed facts, it will largely consist of an elaboration of its initial assumptions about the nature of inheritance and the sources of heritable variation.

The orthodox mechanistic interpretation is provided by the neo-Darwinian theory, which differs from the original Darwinian theory in two major respects: first, it assumes that heredity is explicable in terms of genes and chromosomes; and second, that the ultimate source of heritable variability is the random mutation of the genetic material. The main features of this theory can be summarised as follows:

(i) Mutations take place at random.

(ii) Genes are recombined by sexual reproduction, the 'crossing over' of chromosomes, and by changes in chromosomal structure. These processes produce new permutations of genes which may bring about new effects.

(iii) The spread of a favourable mutation is likely to be more rapid in small than in large interbreeding populations. In small populations or in medium-sized populations which undergo large fluctuations, mutant genes may be lost or preserved at random by 'genetic drift' rather than as a result of natural selection.

(iv) Natural selection tends to eliminate mutant genes with harmful effects. The agents of selection include predators; parasites and infectious diseases; competition for space, food, etc.; climatic and micro-climatic conditions; and sexual selection.

(v) New selection pressures come into play as a result of changes in environmental conditions, and of changes in the behavioural patterns of the organisms themselves.

(vi) If populations become separated geographically or ecologically, or for any other reason, they are likely to undergo divergent evolution.

(vii) Particularly in the plant kingdom, new species may arise from inter-specific hybrids which, although usually sterile, sometimes become fertile as a result of polyploidy.

Some of the main features of this neo-Darwinian theory have been elaborated mathematically in the field of theoretical population genetics. In order to construct mathematical models, it is usually assumed for the sake of simplicity that genes are subjected to selection independently of each other (although in fact they are linked together in chromosomes and interact in their effects with other genes). Then by assigning numerical values to selection pressures, mutation rates and population sizes, the changes in gene frequency over a given number of generations can be worked out. These methods have been extended to cover all aspects of evolution by assuming that morphological characters and instincts are determined by individual genes or combinations of genes.[1]

Most neo-Darwinian theorists assume that divergent evolution under the influence of natural selection over long periods of time will not only lead to the development of new races, varieties and sub-species, but also new species, genera, families, orders and phyla.[2] This view has been disputed on the grounds that the differences between these higher taxonomic divisions are too great to have arisen by gradual transformations; apart from anything else, the organisms often differ in the number and structure of their chromosomes. Several authors have suggested that these large-scale evolutionary changes occur suddenly as a result of macro-mutations. Contemporary examples of such sudden changes are provided by monstrous animals and plants in which structures have been transformed, reduplicated or suppressed. Occasionally in the course of evolution, 'hopeful monsters' could have survived and reproduced.[3] One argument advanced in favour of this view is that whereas gradual changes under selection pressure should result in forms with a definite adaptive value (except perhaps in small populations subject to 'genetic drift'), macro-mutations could produce all sorts of apparently gratuitous large-scale variations which would be weeded out by natural selection only if they were positively harmful, thus helping to account for the prodigious diversity of living organisms.[4]

Although these authors emphasize the importance of sudden large changes, they do not disagree with the orthodox assumptions that evolution as a whole depends only on random mutations and genetical inheritance, in combination with natural selection.

More radical critics challenge these basic principles themselves, arguing that it is hardly conceivable that all the adaptive structures and instincts of living organisms could have arisen purely by chance, even granted that natural selection will only permit organisms to survive and reproduce if they are sufficiently well-adapted to do so. Moreover, they claim that some instances of parallel and convergent evolution, in which very similar morphological characters appear independently in different taxonomic groups, indicate the operation of unknown factors in evolution, even granted parallel selection pressures. Finally, some object to the implicit or explicit mechanistic assumption that evolution as a whole is entirely purposeless.[5]

The metaphysical denial of any creative agency or purpose in the evolutionary process follows from the philosophy of materialism, with which the mechanistic theory is so closely associated.[6] But unless scientific and metaphysical issues are to become hopelessly confounded, within the limited context of empirical science the neo-Darwinian theory must be treated not as a metaphysical dogma, but as a scientific hypothesis. As such it can hardly be regarded as proved: at best it offers a plausible interpretation of the processes of evolution on the basis of its assumptions about genetical inheritance and the randomness of mutations.

The hypothesis of formative causation enables heredity to be seen in a new light, and therefore leads to a rather different interpretation of evolution. But while agreeing with the neo-Darwinian assumption that genetic mutations are random, it neither affirms nor denies the metaphysics of materialism (Section 8.7).

8.2 Mutations

If organisms developed in the same environments generation after generation, and passed on identical genes and chromosomes to their offspring, the combined effects of genetic inheritance and morphic resonance would lead to an indefinite repetition of the same old forms. But, in fact, changes are imposed upon organisms both from within, by genetic mutation, and from without, by alterations in the environment.

Mutations are accidental changes in the structure of genes or of chromosomes, individually unpredictable not only in practice, but also in principle, because they depend on probabilistic events. There seems no reason to doubt that they are random, as the neo-Darwinian theory supposes.

Many mutations have effects which are so deleterious as to be lethal. But of those which are less harmful, some affect morphogenesis through quantitative influences on pathways of morphogenesis, and give rise to variants of normal forms (Section 7.3); and others affect morphogenetic germs in such a way that whole pathways of morphogenesis are blocked, or replaced by other pathways (Section 7.2).

In those rare cases where mutations lead to changes which are favoured by natural selection, not only will the proportion of mutant genes in the population tend to increase, in accordance with the neo-Darwinian theory, but also the repetition of the new pathways of morphogenesis in increasing numbers of organisms will reinforce the new chreodes: not only the 'gene pools', but also the morphogenetic fields of a species will change and evolve as a result of natural selection.

8.3: The divergence of chreodes

If a mutation or environmental change perturbs a normal pathway of morphogenesis at a relatively early stage, the system may be able to regulate and go on to produce a normal final form in spite of this disturbance. If this process is repeated generation after generation, the chreodic diversion will be stabilized by morphic resonance; consequently a whole race or variety of a species will come to follow an abnormal pathway of morphogenesis while still ending up with the usual adult form. In fact many cases of so-called temporary deviations in development have been described. For example, in the turbellarian worm *Prorhynchnus stagnitilis* the egg cells cleave either in a spiral or in a radial manner, and the developing embryos grow either inside the yolk or on its surface. Owing to these differences in early embryology, some of the organs are formed in different sequences; nevertheless, the adult animals are identical. And in a single species of the annelid worm *Nereis*, two very different kinds of larva are produced; but both develop into the same adult form.[7] In some such cases, the temporary deviations may be adaptive, for example to conditions of larval life, but in most they occur for no apparent reason.

Of much greater evolutionary significance are those divergences of chreodes which are not fully corrected by regulation and which therefore give rise to variant final forms. Such changes in the pathway of development could arise either as a result of mutations (cf. Section 7.3) or unusual environmental conditions (cf. Section 7.6). In the case of mutation in an unchanged environment, if the deviant final form has a selective advantage, the mutant genes will increase in frequency within the population, and also the new chreode will be increasingly reinforced by morphic resonance. In the more complicated case where a variant form which arises in

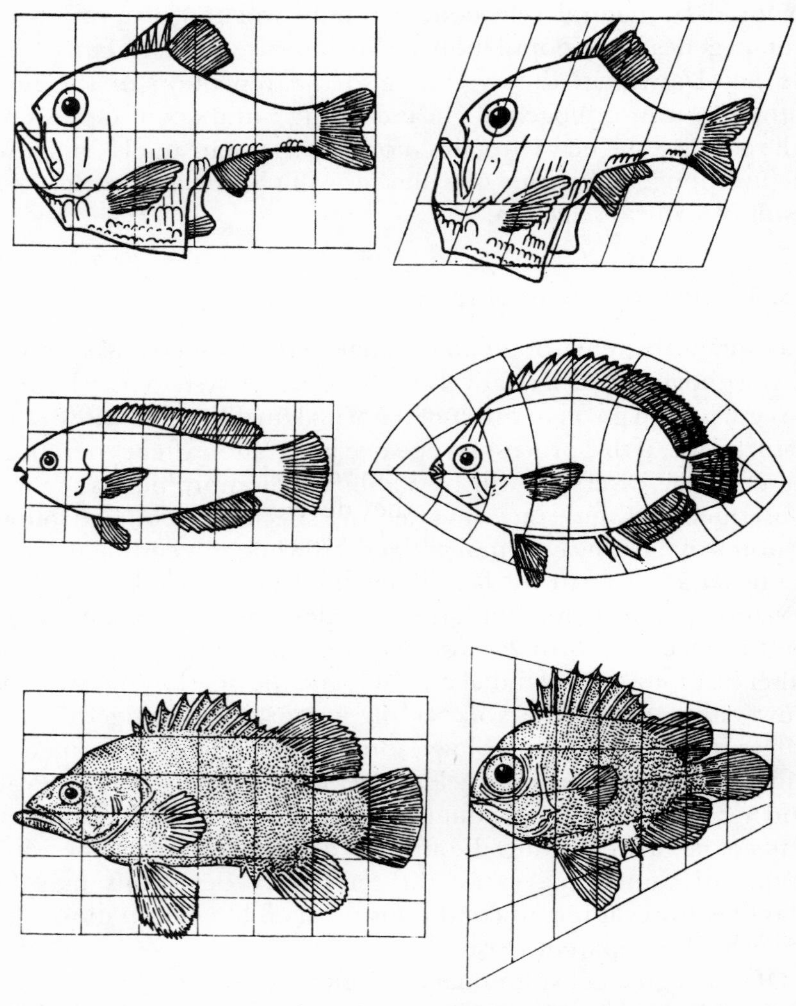

Figure 20 Comparisons of the forms of different species of fish. (From Thompson, 1942. Reproduced by courtesy of the Cambridge University Press).

response to unusual environmental conditions has a selective advantage, the new chreode will be reinforced as before, and at the same time selection will also operate in favour of those organisms with the genetic capacity to respond in this way (cf. Waddington's experiments on fruit flies, Section 7.7). So the acquired characteristics will become hereditary through a combination of genetic selection and morphic resonance.

Under natural conditions, the operation of different selection pressures on geographically or ecologically isolated populations of a species will result in a divergence both of their 'gene pools' and of their chreodes. Countless species of animals and plants have in fact diverged into genetically and morphologically distinct races and varieties; familiar examples are provided by domesticated animals and cultivated plants.[8] Think, for instance, of the amazingly diverse breeds of dog, ranging from the Afghan hound to the Pekinese.

In some cases, the morphological divergence affects only one particular structure or a small group of structures, while others remain relatively unaffected. For example in the small fish *Belone acus* the jaws in the early stages of development resemble those of related species, but subsequently they develop into an enormously elongated snout.[9] Many structural exaggerations have evolved under the influence of sexual selection, for example the antlers of deer. And flowers provide thousands of examples of the divergent development of different component parts: compare, for instance, the modifications of the petals in different species of orchid.

In other cases, the form of many different structures has changed in a correlated manner. Indeed if the forms vary in a particularly uniform and harmonious way, they can be compared diagrammatically by the systematic distortion of superimposed co-ordinates (Fig. 20), as Sir D'Arcy Thompson showed in the chapter of his essay *On Growth and Form* entitled 'The Theory of Transformations, or the Comparison of Related Forms'.

These kinds of evolutionary change take place within the context of already-existing morphogenetic fields. They produce variations on given themes. But they cannot account for these themes themselves. In Thompson's words:

> 'We *cannot* transform an invertebrate into a vertebrate, nor a coelenterate into a worm, by any simple and legitimate deformation, nor by anything short of reduction to elementary principles . . . Formal resemblance, which we depend on as a trusty guide to the affinities of animals within certain bounds or grades of kinship and propinquity, ceases in certain other cases to serve us, because under certain circumstances it ceases to exist. Our geometrical analogies weigh heavily against

Darwin's conception of endless small continuous variations; they help to show that discontinuous variations are a natural thing, that . . . sudden changes, greater or less, are bound to have taken place, and new 'types' to have arisen, now and then'.[10]

8.4 The suppression of chreodes

Whereas the divergence of chreodes within existing morphogenetic fields permits continuous or quantitative variation of form, developmental changes involving the suppression of chreodes or the substitution of one chreode for another result in qualitative discontinuities. According to the hypothesis of formative causation, these effects are caused by mutations or environmental factors which alter morphogenetic germs (Section 7.2). Examples of mutant pea leaves in which leaflets are substituted for tendrils are shown in Fig. 18, and a 'bithorax' mutant of *Drosophila* in Fig. 17.

Changes of these types probably occurred frequently in the course of evolution. For example, in certain species of *Acacia*, the leaves have been suppressed and their role taken over by flattened leaf-stalks. This process can actually be seen in seedlings, where the first-formed leaves are typically pinnate (Fig. 21). In members of the cactus family, leaves have been replaced by spines. Among the insects, in almost every order there are species in which the wings have been suppressed either in both sexes, as in certain parasitic flies, or in only one sex, as in the female beetle known as the glow worm. In the case of ants, female larvae develop either into winged queens or wingless workers depending on the chemical constitution of their diet.

In some species, juvenile forms become sexually mature and reproduce without ever producing the characteristic structures of the adult, which are, as it were, short-circuited. The classic example is the axolotl, a tadpole of the tiger salamander, which reaches full size and becomes sexually mature without losing its larval characteristics. If axolotls are supplied with thyroid hormone, they metamorphose into the air-breathing adult form and move out of the water onto land.

The most extreme examples of the suppression of chreodes are found among parasites, some of which have lost nearly all the

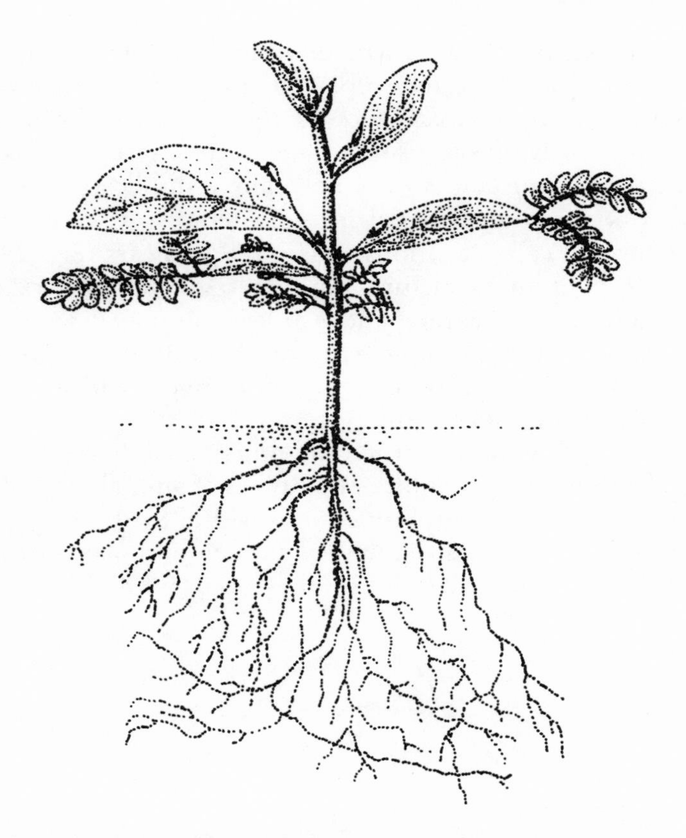

Figure 21 A seedling of an *Acacia* species. (After Goebel, 1898).

structures characteristic of related free-living forms.

8.5 The repetition of chreodes

In all multicellular organisms, some structures are repeated several or many times: the tentacles of *Hydra*, the arms of the starfish, the legs of centipedes, the feathers of birds, the leaves of trees, and so on. Then many organs are made up repeated structural units: the tubules of kidneys, the segments of fruits, etc. And, of course, at the microscopic level, tissues contain thousands or millions of copies of a few basic types of cell.

If, as a result of mutations or environmental changes, extra morphogenetic germs are formed within developing organisms, then certain structures can be repeated more than usual. A familiar horticultural example is that of 'double' flowers, containing additional petals. Human babies are sometimes born with extra fingers or toes. And many instances of abnormally reduplicated structures can be found in the standard texts on teratology, ranging from double-headed calves to monstrous multiple pears (Fig. 22).

As these additional structures develop, regulation occurs in such a way that they are integrated more or less completely with the rest of the organism: for example extra petals in double flowers have normal vascular connections, and extra fingers and toes have a proper blood supply and innervation.

That reduplication of structural units must have played an essential role in the evolution of new types of animals and plants is evident from the structural repetitions within existing organisms. Moreover, many of the structures of animals and plants which are now different from each other may well have evolved from

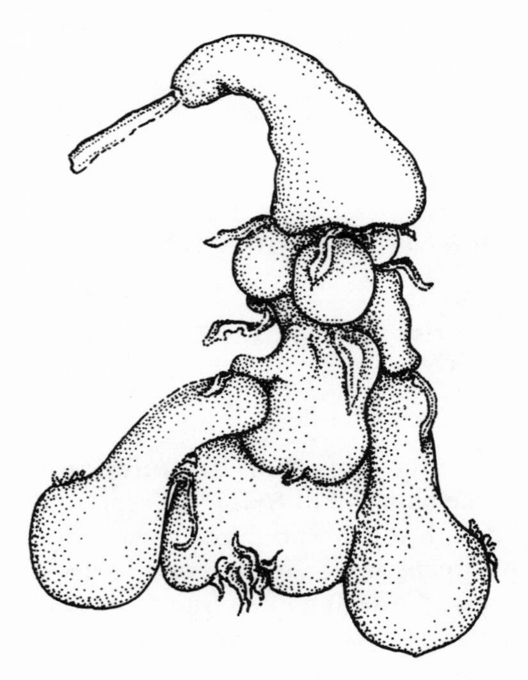

Figure 22 A monstrous pear. (After Masters, 1869).

originally similar units. For example, the insects are believed to have evolved from creatures resembling primitive centipedes, with a series of more or less identical segments, each bearing a pair of leg-like appendages. The appendages on the segments at the front end may have given rise to the mouthparts and antennae, while the segments themselves fused together to form the head. At the tail end some of the appendages may have been modified to produce structures concerned with mating and manipulation of the eggs. In the abdominal segments the appendages were suppressed, but in the three thoracic segments they were retained, and evolved into the modern insect legs.[11]

Such a divergence of originally similar chreodes would only have been possible if the segmental morphogenetic germs became differentiated from one another in their structure; otherwise they would all have continued to become associated by morphic resonance with the same morphogenetic fields. And even in modern insects, if this diversification of the segmental primordia failed to occur during the early stages of embryology, then the normal differences between segments would be lost. Indeed, this is just what seems to happen in the fruit fly *Drosophila* as a result of mutations in the 'bithorax' gene complex: some transform the structures of the third thoracic segment into those of the second, so the fly bears two pairs of wings instead of one (Fig. 17); some transform abdominal segments into thoracic type segments, bearing legs; and others have the reverse effect, transforming thoracic segments into ones of the abdominal type.[12]

8.6 The influence of other species

Practical breeders of animals and plants noticed long ago that cultivated varieties from time to time produced offspring resembling the ancestral wild type. Moreover when two distinct cultivated varieties were crossed, the characters of the offspring sometimes resembled neither of the parental types, but rather those of the wild ancestors. This phenomenon was referred to as 'reversion' or 'atavism'.[13]

In an evolutionary context certain kinds of morphological abnormality can likewise be thought of as reversions to patterns of development of more or less remote ancestral species. For instance, the abnormal formation of two pairs of wings in 'bithorax' mutants of *Drosophila* (Fig. 17) has been interpreted as a 'throw back' to the type of development characteristic of the four-winged ancestors of the flies.[14] Many more examples of putative atavisms can be found in the teratological literature.[15] Of course such interpretations can only be speculative, but they are not necessarily far-fetched. Mutations or abnormal environmental factors could give rise to internal conditions within embryonic tissues which resembled those in ancestral types, with similar morphogenetic consequences.

In most plants and animals, only a small proportion, perhaps less than five percent, of the chromosomal DNA contains genes coding for the organisms' proteins. The function of the great majority of the DNA is unknown. Some may play a part in the control of protein synthesis; some may have a structural role in the chromosomes; and some may consist of 'redundant' ancestral genes which are no longer expressed. It has been suggested that if a mutation – for example due to a rearrangement of chromosome structure – led to the expression of such 'latent' genes, proteins characteristic of remote ancestors might suddenly be produced again, in some cases resulting in the reappearance of long-lost structures.[16]

In terms of the hypothesis of formative causation, if any such changes caused a morphogenetic germ to take up a structure and vibrational pattern similar to that of an ancestral species, it would come under the influence of a morphogenetic field of this species, even though it may have been extinct for millions of years. Moreover, this effect need not be confined to ancestral types. If as a result of mutation (or for any other reason) a germ structure in a developing organism became sufficiently similar to a morphogenetic germ in any other species, contemporary or extinct, it would 'tune in' to a chreode characteristic of this other species. And if the cells were capable of synthesizing appropriate proteins, the system would then actually develop under its influence.

In the course of evolution, closely similar structures sometimes seem to have appeared quite independently in more or less distantly related lines. For example among the Mediterranean dry-land snails, species belonging to well-differentiated genera, identifiable by their genitalia, have shells of nearly identical shape and structure; genera of fossil ammonites show the repeated parallel development of keeled and grooved shells; and similar or identical wing patterns occur in quite different families of butterfly.[17]

If a mutation resulted in an organism 'tuning in' to another species' chreodes and consequently developing structures characteristic of that other species, it would soon be eliminated by natural selection if these structures reduced its chances of survival. On the other hand, if it were favoured by natural selection, the proportion of such organisms in the population would tend to increase. Indeed the selection pressures which favoured its increase might well resemble those which favoured the original evolution of this particular character in the other species. And sometimes the structural resemblance might even be favoured for its own sake, precisely because it enabled the organism to mimic members of another species. Thus evolutionary parallelisms may often depend both on one species 'picking up' the morphogenetic fields of another, and also on parallel selection pressures.

On the other hand, similar selection pressures could also lead to the convergent evolution of superficially similar structures in different species through the modification of different morphogenetic fields. But in such cases, unless the structures closely resembled each other in internal detail as well as external shape, they would be unlikely to interact by morphic resonance.

8.7 The origin of new forms

According to the hypothesis of formative causation, morphic resonance and genetic inheritance together account for the repetition of characteristic patterns of morphogenesis in successive generations of plants and animals. Moreover, characteristics acquired in response to the environment can become hereditary through a combination of morphic resonance and genetic selection. The morphology of organisms can be changed through the suppression or repetition of chreodes; and some striking instances of parallel evolution can be attributed to the 'transfer' of chreodes from one species to another.

However, neither the repetition, modification, addition, subtraction nor permutation of existing morphogenetic fields can explain the origin of these fields themselves. Nevertheless, during the course of evolution, entirely new morphic units together with their morphogenetic fields must have come into being: those of the organelles; of the basic types of cells, tissues and organs; and of the fundamentally different kinds of lower and higher plants and animals.

Although genetic mutations and abnormal environments may well have provided the occasions for the first appearances of new biological morphic units, the forms of their morphogenetic fields could neither have been fully determined by energetic causation, nor by pre-existing formative causes (Section 5.1). It is a matter for conjecture whether any given morphogenetic field originated suddenly in one large 'jump' or more gradually through a series of smaller 'jumps'. But in either case, the new forms taken up in these 'jumps' cannot be explained from within the framework of science in terms of preceding causes.

The origin of new forms could be ascribed either to the creative activity of an agency pervading and transcending nature; or to a creative impetus immanent in nature; or to blind and purposeless chance. But a choice between these metaphysical possibilities could never be made on the basis of any empirically testable scientific hypothesis. Therefore from the point of view of natural science, the question of evolutionary creativity can only be left open.

Notes

1 See for example Wilson (1975).
2 Comprehensive statements of the neo-Darwinian position can be found in Huxley (1942); Rensch (1959); Mayr (1963); and Stebbins (1974).
3 Goldschmidt (1940); Gould (1980).
4 This argument is put forward with many examples by Willis (1940).
5 Perhaps the most stimulating critique of the mechanistic theory of evolution is still H. Bergson's *Creative Evolution* (1911). Bergson does not argue that evolution as a whole has a purpose and direction. This case is put by P. Teilhard de Chardin (1959). For a recent discussion, see Thorpe (1978).
6 See for example Monod (1972).
7 Rensch (1959).
8 For many instructive examples, see Darwin (1875).
9 Rensch (1959).
10 Thompson (1942), pp.1094-1095.
11 Wigglesworth (1964).
12 Lewis (1963, 1978).

13 See the chapter entitled 'Reversion or Atavism' in Darwin (1875).
14 Lewis (1978).
15 E.g. Penzig (1922). For recent discussions see Dostal (1967) and Riedl (1978).
16 R.J. Britten in Duncan and Weston-Smith (eds.) (1977).
17 Rensch (1959).

9 Movements and Motor Fields

9.1 Introduction

The discussion in the preceding chapters concerned the role of formative causation in morphogenesis. The subject of this and the following two chapters is the role of formative causation in the control of movement.

Some of the movements of plants and animals are spontaneous; that is to say they take place in the absence of any particular stimulus from the environment. Other movements take place in response to environmental stimuli. Of course, organisms respond passively to gross physical forces – a tree may be blown over by the wind, or an animal may be carried away by a strong current of water – but many responses are active, and cannot be explained as gross physical or chemical effects of the stimuli on the organism as a whole: they reveal the organism's *sensitivity* to the environment. This sensitivity generally depends on specialized receptors or sense organs.

The physico-chemical basis of the excitation of these specialized receptors by stimuli from the environment has been worked out in considerable detail; so has the physiology of the nerve impulse; and so has the functioning of the muscles and other motor structures. But very little is known about the control and co-ordination of behaviour.

In this chapter it is suggested that just as formative causation organizes morphogenesis through the probability structures of fields which impose pattern and order on energetically indeterminate processes, so it organizes movements, and hence behaviour. The similarities between morphogenesis and behaviour are not immediately obvious, but are easiest to understand in the case of plants and unicellular animals such as *Amoeba*, whose movements are essentially morphogenetic. These are considered first.

9.2 The movements of plants

Plants generally move by growing.[1] This fact becomes easier to appreciate when they are seen on speeded-up films: shoots stretch out and curve towards the light; tap roots thrust downwards into the soil; and the tips of tendrils and climbing stems sweep out wide spirals in the air until they make contact with a solid support and coil around it.[2]

The growth and development of plants takes place under the control of their morphogenetic fields, which give them their characteristic forms. But the orientation of this growth is determined to a large extent by the directional stimuli of gravity and light. Environmental factors also influence the type of development: for example in dim light plants become etiolated; their shoots grow relatively rapidly in a spindly manner until they get into brighter light.

Gravity is 'sensed' through its effects on starch grains, which roll downwards and accumulate in the lowest parts of the cells.[3] The direction from which light is coming is detected by the differential absorption of radiant energy on the illuminated and shaded sides of organs by a yellow carotenoid pigment.[4] The sense of 'touch' by which climbing shoots and tendrils locate solid supports may involve the release of a simple chemical, ethylene, from the surface cells which are mechanically stimulated.[5] The change-over from etiolated to normal growth depends on the absorption of light by a blue protein pigment called phytochrome.[6]

The responses to these stimuli involve complicated physico-chemical changes within the cells and tissues, and in some cases depend on the differential distribution of hormones such as auxin. However, the reactions cannot be explained in terms of these physico-chemical changes alone, but can only be understood within the context of the plants' overall morphogenetic fields. For example, owing to their inherent polarity, plants produce shoots at one end and roots at the other. The directional stimulus of gravity orients this polarized development so that the shoots grow upwards and the roots downwards. The action of the gravitational field on starch grains within the cells and consequent changes in hormonal distribution are indeed causes of these oriented growth movements, but cannot in themselves account for the pre-existing polarity; nor for the fact that the main shoots and roots respond in exactly opposite senses; nor for the different habits of growth of trees, herbs, climbers and creepers; nor for the particular patterns of branching in the shoot and root systems of different species. All these characteristics depend on the morphogenetic fields.

Although most of the movements of plants occur only in young growing organs, some structures retain the ability to move even when they are mature, for example flowers which open and close again daily, and leaves which fold up at night. These movements are influenced by the intensity of the light and other environmental factors; they are also under the control of a 'physiological clock' and continue to take place at approximately daily intervals even if the plants are placed in an unchanging environment.[7] The leaves or petals open up because specialized cells in the 'hinge' region at their bases become turgid; they close when these cells lose water owing to changes in the permeability of their membranes to inorganic ions.[8] The regaining of turgor is an active, energy-requiring process, comparable to growth.

In addition to making 'sleep' movements, the leaves of some species move during the course of the day in response to the changing position of the sun. For example, in the pigeonpea, *Cajanus cajan*, the leaflets exposed to the sun are oriented approximately parallel to the sun's rays, exposing the minimum surface area to the intense tropical radiation. But leaves in the shade orient themselves at right angles to the incident radiation, thus intercepting the maximum amount of light. These responses depend on the direction and intensity of light falling on the specialized leaf-joints, the pulvini. Throughout the day the leaves and leaflets are continuously adjusting their positions as the sun moves across the sky. At night they take up their vertical 'sleep' positions: the pulvini are sensitive to gravity as well as light.

In the 'sensitive plant', *Mimosa pudica*, the leaflets close up and the leaves point downwards at night, as they do in many other leguminous plants. But these movements also occur rapidly during the daytime in response to mechanical stimulation (Fig. 23). The stimulus causes a wave of electrical depolarization, similar to a nerve impulse, to pass down the leaf; if the stimulus is strong

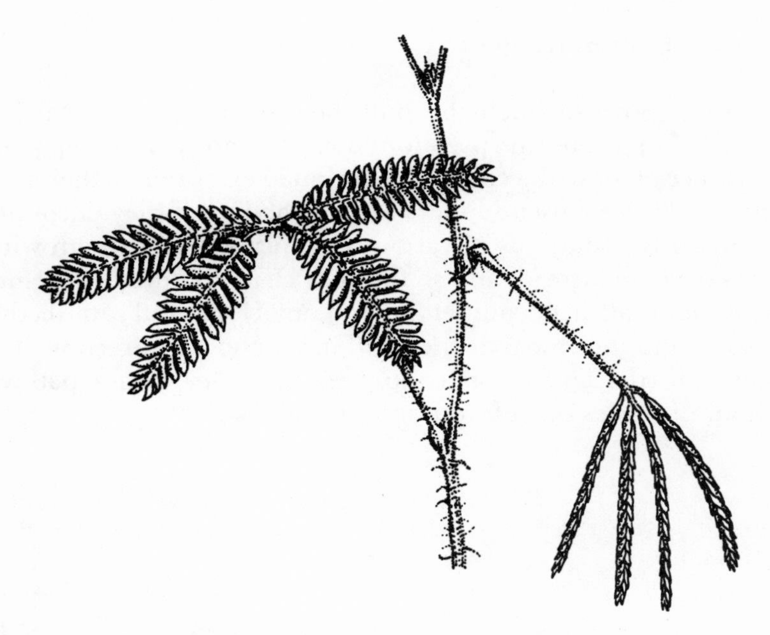

Figure 23 Leaves of the sensitive plant, *Mimosa pudica*. Left, unstimulated; right, stimulated.

enough, the impulse spreads to other leaves, which also fold up.[9] Similarly, in the Venus' fly trap, *Dionaea muscipula*, mechanical stimulation of the sensitive hairs on the surface of the leaf causes an electrical impulse to travel to the turgid 'hinge' cells, which rapidly lose water; the leaf closes like a trap around hapless insects, which are then digested.[10]

These movements of leaves and leaflets in response to light, gravity and mechanical stimulation are made possible by the fact that specialized cells are able to lose water and then 'grow' again; they consequently retain a simplified morphogenetic potential, while that of most other tissues is lost when they mature and cease to grow. The reversible movements of these specialized structures can be regarded as limiting cases of morphogenesis in which the changes of form have become stereotyped and repetitive. But their quasi-mechanistic simplicity is evolutionarily secondary, not primary; it has evolved from a background in which sensitivity to environmental stimuli is associated with the growth and morphogenesis of the plant as a whole.

9.3 Amoeboid movement

Amoebae move through the bulk flow of their cytoplasm into growing projections, the pseudopodia. They normally creep along the surface of solid objects by the continued extension of their front ends. But if these pseudopodia are touched, or if they encounter heat or strong solutions of various chemicals, they stop growing; others develop instead, and so the cells change course. If the new pseudopodia again encounter any potentially harmful stimuli, they too stop, and the amoebae move off in yet another direction. This system of 'trial and error' continues until they find a pathway without obstacles or unfavourable stimuli.[11]

Figure 24 Method by which a floating amoeba passes to a solid surface. (After Jennings, 1906).

In free-floating amoebae not exposed to any particular directional stimulus, there is no consistent orientation of growth; pseudopodia keep developing in various directions until one of them comes into contact with a surface along which it can creep (Fig.24).

The extension of pseudopodia presumably occurs under the

influence of a specific polarized morphogenetic field. The orientation in which new pseudopodia start to form may depend to a large extent on chance fluctuations within the cell; the virtual pseudopodia projected outwards from the cell-body are then actualized through the organization of contractile filaments and other structures within the cytoplasm. This process continues until the development of the pseudopodia is inhibited by stimuli from the environment, or by competition from pseudopodia growing in other directions.

The fact that amoeboid movements depend on continuous morphogenetic processes is aptly indicated in the specific name of *Amoeba proteus* by the allusion to the mythical sea-deity who kept changing from one shape to another.

Amoebae feed by engulfing food particles, such as bacteria, by the process of phagocytosis: pseudopodia grow around the particle which is in contact with the surface of the cell; the membranes of the pseudopodia fuse together, and the particle is enclosed within the cell surrounded by a part of the cell membrane. Other membrane-bound vesicles containing digestive enzymes fuse with this phagocytotic vesicle and the food is digested. This type of morphogenesis is distinct from that of cellular locomotion and presumably takes place under the influence of a different morphogenetic field, the orientation of which depends on the contact of the potential food particle with the membrane. This particle in contact with the membrane can be regarded as the morphogenetic germ; the final form is the particle engulfed within the cell. The chreode of phagocytosis leading to this final form is given by morphic resonance from all similar acts of phagocytosis by similar amoebae in the past.

9.4 The repetitive morphogenesis of specialized structures

The movements of most animals depend on the change of form of certain specialized structures, rather than of the body as a whole.

Many unicellular organisms are propelled by the beating of whip-like outgrowths, the flagella or cilia, while the form of the rest of the cell remains more or less fixed (Fig. 25). These motile organelles contain long tubular elements very similar to cytoplasmic microtubles; the change of shape of proteins associated with the tubules generates a sliding shear force, resulting in the bending of the flagella or cilia.[12]

In ciliates, the movements of the many individual cilia are co-ordinated in such a way that waves of beating pass over the surface of the cell. In some species, this co-ordination seems to depend on the mechanical influence of the cilia on their neighbours; and in others, on an excitatory system within the cell probably associated with fine fibrils connecting together the bases of the cilia.[13]

If a swimming ciliate, for example *Paramecium*, meets with an unfavourable stimulus, the direction of ciliary beating is reversed: the organism backs away and then swims forward again in a new direction.[14] This avoidance reaction is probably triggered by the entry of calcium or other ions into the cell as a result of an alteration in membrane permeability brought about by the stimulus.[15]

The change of form of the beating flagella and cilia, as well as the control of this beating, takes place in such a stereotyped, repetitive way that it seems almost machine-like.

This quasi-mechanistic specialization of structure and function is taken still further in the multicellular animals. Entire cells and groups of cells are specialized to undergo a repeated, simplified morphogenesis in their cycles of contraction and relaxation; others have a specialized sensitivity to light, chemicals, pressure, vibration or other stimuli; and the nerves, with their enormously elongated axons, are specialized to conduct electrical impulses from place to place, linking the sense organs and the muscles to the nerve net or central nervous system.

9.5 Nervous systems

Just as the beating of the individual cilia on the surface of a ciliate is co-ordinated with that of neighbouring cilia through definite physical connections, so the contraction of individual muscle cells is co-ordinated by means of deterministic impulses passing through the nerves. When several neighbouring cells are activated by a single nerve, they can be caused to contract simultaneously. And

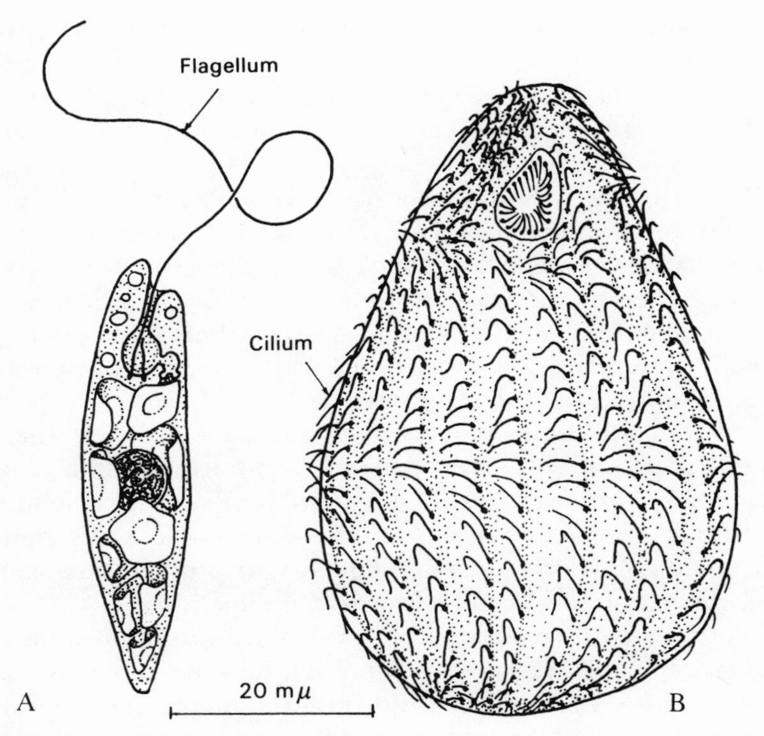

Figure 25 A: A flagellate, *Euglena gracilis.* (After Raven *et al.*, 1976). B: A ciliate, *Tetrahymena pyriformis.* (After Mackinnon and Hawes, 1961).

when the activity of this nerve is part of a higher-level system of control, the contraction of different groups of cells can be co-ordinated in a rhythmical manner, as it is in a muscle which maintains its tension over a period of time. Then yet higher-level systems control repetitive cycles of contraction in different muscles, for example in the legs of an animal as it runs. Thus the hierarchically organized activities of the nervous system permit degrees of co-ordination which would be impossible if the fields controlling the movements of organisms acted directly on the muscle cells.

But although on the one hand nerves function deterministically in the transmission of definite 'all or none' impulses from one place to another, on the other hand formative causation would not be able to control animals' movements through the nervous system unless the activity of the nerves was at the same time inherently probabilistic. And in fact it is.

The firing of nerve impulses depends on changes in the permeability of the membranes of nerve cells to inorganic ions, in particular sodium and potassium. These changes can be brought about either by electrical stimulation, or by specific chemical transmitters (e.g. acetylcholine) released from nerve endings at synaptic junctions (Fig. 26). The excitation of nerves by electrical stimuli around the threshold level has long been known to take place probabilistically.[16] The main reason for this is that the electrical potential across the membrane fluctuates in a random manner.[17] Moreover, the changes in post-synaptic membrane potentials caused by chemical transmitters also show random fluctuations,[18] which seem to be due to the probabilistic opening and closing of ionic 'channels' across the membrane.[19]

Not only is there an inherent probabilism in the responses of post-synaptic membranes to chemical transmitters, but also in the release of the transmitters from the pre-synaptic nerve endings. Transmitter molecules are stored in numerous microscopic vesicles (Fig. 26), and are released into the synaptic cleft when these vesicles fuse with the membrane. This process occurs spontaneously at random intervals, giving rise to discharges of so-called miniature end-plate potentials. The rate of secretion is greatly increased when an impulse arrives at the nerve ending, but here again the fusion of the vesicles with the membrane takes place probabilistically.[20]

Within the brain, a typical nerve cell has thousands of fine thread-like projections that end in synaptic junctions on other nerve cells; and, conversely, projections from hundreds or thousands of other nerve cells end in synapses on its own surface (Fig. 26). Some of these nerve endings release excitatory transmitters that tend to promote the firing of an impulse; others are inhibitory and reduce the tendency of the nerve to fire. The triggering of impulses in fact depends on a balance of excitatory and inhibitory influences from hundreds of synapses. It seems likely that, at any given time, in many of the nerve cells in the brain this balance is poised so critically that firing either occurs or does not occur as a result of

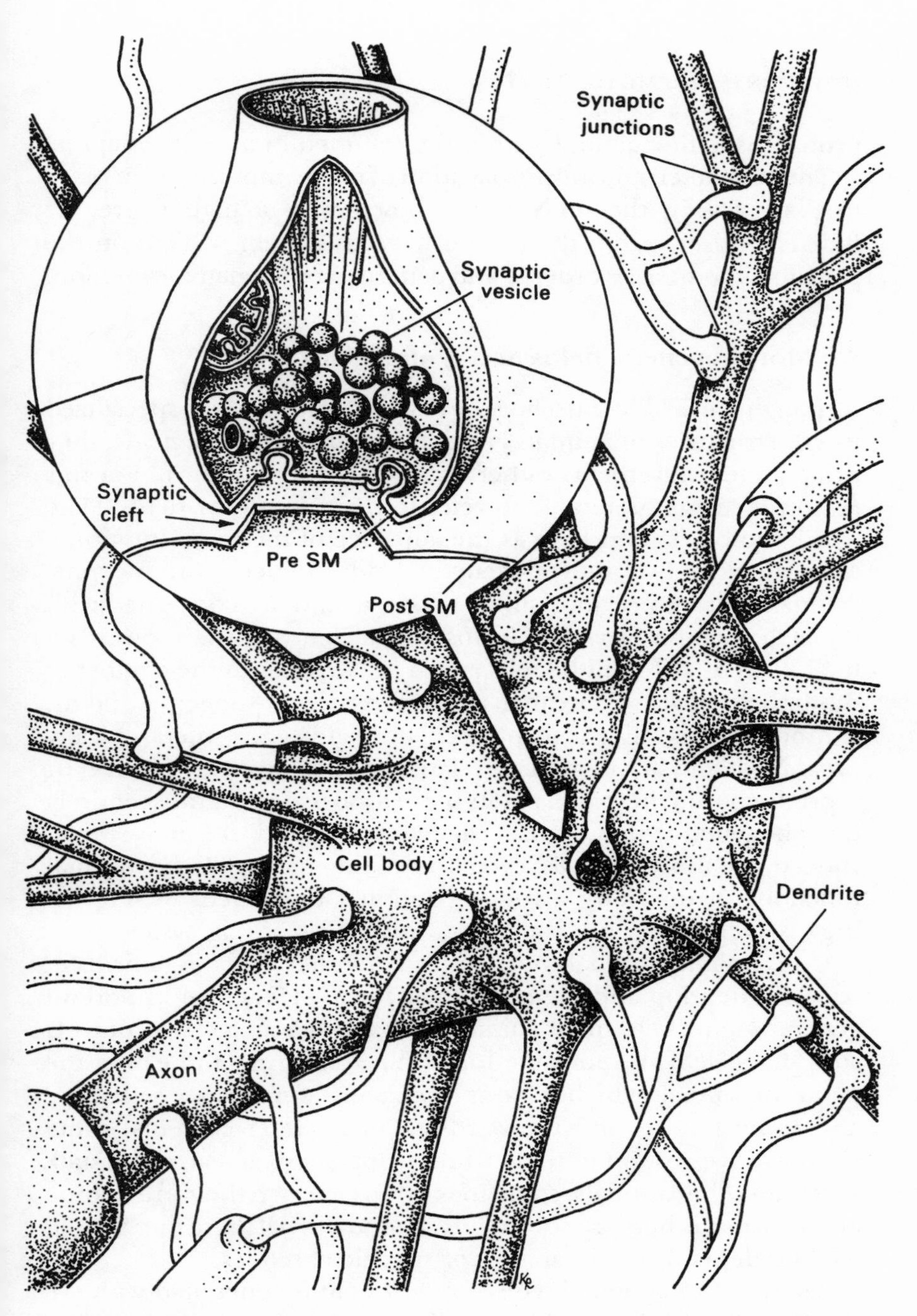

Figure 26 Part of a nerve cell, with numerous synapses on its surface. The inset shows an individual synapse in more detail. Pre SM=pre-synaptic membrane; Post SM=post-synaptic membrane. (Based on Krstić, 1979).

probabilistic fluctuations within the cell membranes or synapses.

Thus the deterministic propagation of nerve impulses from place to place within the body is combined with a high degree of indeterminism within the central nervous system, which, on the present hypothesis, is ordered and patterned by formative causation.

9.6 Morphogenetic fields and motor fields

Although the fields controlling the changes of form of the specialized motor structures of animals are in fact morphogenetic fields, they bring about movements rather than net changes of form. For this reason it seems preferable to refer to them as motor fields. (The word 'motor' is used here as the adjective of the noun 'motion'.) Motor fields, like morphogenetic fields, depend on morphic resonance from previous similar systems and are concerned with the actualization of virtual forms. Canalized pathways towards a final form or state can be referred to as chreodes in the context of motor fields just as they can in the context of morphogenetic fields.

Motor fields, like morphogenetic fields, are hierarchically organized, and are in general related to development, survival or reproduction. Whereas in plants these processes are almost entirely morphogenetic, in animals they also depend on movement. Indeed, in most animals even the maintenance of the normal functions of the body involves continual movement of internal organs such as the gut, the heart, and the breathing system.

Unlike plants, animals need to feed on other living organisms in order to develop and maintain their forms. Hence an important general motor field in all animals is that of feeding. This controls subsidiary fields responsible for finding, securing and eating the plants or animals which serve as food. Some animals are sedentary and cause food to move towards them in water currents; some simply move around until they find plants they can eat; some stalk and hunt other animals; some make traps to catch their prey; some are parasitic; others are scavengers; and so on. All these methods of feeding depend on hierarchies of specific chreodes.

Another fundamental type of motor field is concerned with the avoidance of unfavourable conditions. *Amoeba* and *Paramecium* show the simplest type of reaction: backing or turning away from the unfavourable stimulus and setting off in some other direction.

Sedentary animals such as *Stentor* and *Hydra* react to mildly unfavourable stimuli by contracting their bodies, but in response to more severe stimuli they move away and settle down somewhere else. In addition to general avoidance reactions, many animals also exhibit special types of behaviour which help them to escape from predators; for example they may run away swiftly, or stand their ground and somehow frighten the predator, or 'freeze' in such a way that they are less easily seen.

The overall fields of development and survival have as their final form the fully-grown animal under optimal conditions. Whenever this state is reached, there is no need for the animal to do anything in particular; but deviations from this state bring the animal under the influence of the various motor fields directed towards its restoration. In fact, such deviations are frequent: the animal's continuous metabolism depletes its reserves of food; changes in the environment expose it to unfavourable conditions; and predators approach it unpredictably. These and other changes are detected by the sensory structures and result in characteristic modifications of the nervous system, which then becomes the germ structure for particular motor fields.

The final form of the overall field of reproduction is the establishment of viable progeny. In unicellular organisms, and in simple multicellular animals such as *Hydra* this is achieved by a morphogenetic process: the organisms divide into two, or 'bud off' new individuals. Likewise, primitive methods of sexual reproduction are essentially morphogenetic: many lower animals (e.g. sea urchins) as well as lower plants (e.g. the seaweed *Fucus*), simply release millions of ova and sperm cells into the water around them.

In more advanced animals, the sperms are released not at random, but in the vicinity of the ova, as a result of specialized mating behaviour. Thus the overall field of reproduction comes to cover the motor fields of searching for a mate, of courtship, and of copulation. Organisms may come under the influence of the first motor field in the sequence as a result of internal physiological changes mediated by hormones, as well as olfactory, visual or other stimuli from potential mates. The end point of the first field constitutes the germ for the second, and so on: searching for a mate is followed by courtship which, if successful, leads to the starting point of the copulation chreode. In the simplest cases, the final form of the whole sequence is for the male ejaculation, and for the female the laying of eggs. In many aquatic organisms they are simply released into the water, but in land animals the deposition of eggs often involves complex and highly specific patterns of behaviour; for example, ichneumon flies inject their eggs into caterpillars of definite species, inside which the larvae develop parasitically, and potter wasps make small 'pots' in which they place paralysed prey before laying their eggs upon the prey and sealing the 'pots'.

In some viviparous species the young are simply released and abandoned at birth. But when the young are cared for after they are born or hatched, a new range of motor fields comes into play, still under the overall field of reproduction of the parents, but at the same time serving the field of development and survival of the young. Consequently, the behaviour of the animals takes on a social dimension. In the simplest cases the societies are temporary and disintegrate when the offspring become independent; in others they persist with a consequent increase in the complexity of behaviour. Special motor fields control the various types of communication between individuals, and the differentiated tasks which different individuals perform.

In the extraordinarily complex societies of the termites, ants and social bees and wasps, individuals of similar or identical genetic constitutions perform quite different tasks, and even the same individual may play different roles at different times – for example a young worker bee may first clean the hive, then after a few days act as a brood nurse, then help build the honey combs, then receive and store pollen, then guard the hive, and finally go out foraging.[21] Each of these roles must be covered by a higher-level motor field, which in turn controls the lower-level chreodes involved in the particular specialized tasks. Changes in the insect's nervous system must bring it under the influence of one or other of these higher-level fields by causing it to enter into morphic resonance with previous workers which filled that particular role. Such changes depend to some extent on alterations in the physiology of the insect as it grows older, but they are also strongly influenced by external stimuli: the roles of individuals change in response to disturbances of the hive or society; the whole system regulates.

The higher-level motor fields of feeding, avoidance, reproduction, etc. generally control a series of lower-level fields which act in sequence, the final form of one providing the germ structure for the next. Motor fields still lower in the hierarchy often act in cycles to give repetitive movements, such as those of the legs in walking, the wings in flying, and the jaws in chewing. At the lowest level are those fields concerned with the detailed control of the contraction of the cells within the muscles.

The higher-level motor fields embrace not only the sense organs, the nervous system and the muscles, but also objects *outside* the animal. Consider, for example, the motor field of feeding. The overall process – the capture and ingestion of food – is in fact a special type of aggregative morphogenesis (cf. Section 4.1). The hungry animal is the germ structure and enters into morphic resonance with previous final forms of this motor field, namely similar past animals, including itself, in a well-fed state. In the case of a predator, the achievement of this final form depends on the capture and ingestion of prey. The motor field of capture projects into the space around the animal, and includes within it the virtual form of the prey (cf. Fig. 11). This virtual form is actualized when an entity corresponding sufficiently closely to it nears the predator: the prey is recognized, and the capture chreode initiated. Theoretically, the motor field could affect probabilistic events in any or all of the systems it embraces, including the sense organs, muscles and prey itself. But in most cases its influence seems likely to be confined to the modification of probabilistic events in the central nervous system, directing the movements of the animal towards the achievement of the final form, in this case the capture of the prey.

9.7 Motor fields and the senses

By morphic resonance, an animal comes under the influence of specific motor fields as a result of its characteristic structure and internal patterns of oscillatory activity. These patterns are modified by changes arising within the body of the animal, and by influences from the environment.

If different stimuli brought about the same changes within the animal, then the same motor fields would come into play. This is what seems to happen in unicellular organisms which give the same avoidance reaction to a wide variety of physical and chemical stimuli: probably all of them have similar effects on the physico-chemical state of the cell, for example by modifying the permeability of the cell membrane to calcium or other ions.

In simple multicellular animals with relatively little sensory specialization, the range of reactions to stimuli is not much greater than in unicellular organisms. *Hydra*, for instance, shows the same avoidance reactions to many different physical and chemical stimuli, and responds to objects such as food particles only as a result of mechanical contact. However, as in certain unicellular organisms, its response to solid objects is modified by chemical stimuli. This can be demonstrated by a simple experiment: if small pieces of filter paper are supplied to the tentacles of a hungry *Hydra*, they evoke no reaction; but if they are first soaked in meat-juice, the tentacles carry them towards the mouth, where they are swallowed.[22]

By contrast, animals possessing image-forming eyes can sense objects while they are still some distance away; consequently, the motor fields project far further outwards into the environment; the range and scope of the animals' behaviour is greatly increased. Similarly, the sense of hearing enables distant objects to be detected and hence permits an extension of the spatial range of the motor fields even into regions which cannot be seen. In some animals, most notably bats, this sense has replaced sight as the basis of the extended motor fields. And in a few aquatic species, such as the Mormyrid and Gymnotid electric fish, specialized receptors detect changes in the electric field set up around themselves by pulses from their electric organs; this sense enables them to locate prey and other objects in the muddy tropical rivers in which they live.

As animals move, the sensory stimuli arising both within their bodies and from the environment change as a consequence of their own movements. This continuous feedback plays an essential part in the co-ordination of movements by their motor fields.

Motor fields, like morphogenetic fields, are probability structures which become associated by morphic resonance with physical systems on the basis of their three-dimensional patterns of oscillation. It is therefore of fundamental significance that all sensory inputs are translated into spatio-temporal patterns of activity within the nervous system. In the sense of touch the stimuli act on particular parts of the body, which through specific nervous pathways are 'mapped' within the brain; in vision, images falling on the retina bring about spatially patterned changes in the optic nerves and visual cortex; and although olfactory, gustatory and auditory stimuli are not directly spatial, the nerves they excite through the relevant sense-organs have specific locations, and the impulses travelling along these nerves into the central nervous system set up characteristic patterns of excitation.

Thus particular stimuli and combinations of stimuli have characteristic spatio-temporal effects. These dynamic patterns of activity bring the nervous system into morphic resonance with similar past nervous systems in similar states, and hence under the influence of particular motor fields.

9.8 Regulation and regeneration

Motor fields, like morphogenetic fields, lead the systems under their influence towards characteristic final forms. They usually do so by initiating a series of movements in a definite sequence. The intermediate stages are stabilized by morphic resonance; in other words, they are chreodes. But chreodes simply represent the most probable pathways towards final forms. If the normal pathway is blocked, or if the system is deflected from it for any reason, the same final form may be reached in a different way: the system regulates (Section 4.1). Many, but not all, morphogenetic systems are capable of regulation; and so are motor systems.

Regulation occurs under the influence of motor fields at all levels in the hierarchy: for example if a few muscles or nerves in a dog's leg are damaged, the pattern of contraction in the other muscles adjusts so that the limb functions normally. If the leg is amputated, the movements of the remaining legs change in such a way that the dog can still walk, although with a limp. If parts of its cerebral cortex are damaged, after some time it recovers more or less completely. If it is blinded, its ability to move around gradually improves as it comes to rely more on its remaining senses. And if its normal route towards its home, its food or its puppies is blocked, it changes its habitual sequence of movements until it finds a new way to reach its goal.

The behavioural equivalent of regeneration occurs when the final form of a chreode has been actualized, but is then disrupted: think, for instance, of a cat which has caught a mouse, the end point of the capture chreode. If the mouse escapes from its clutches, then the cat's movements are directed towards re-capturing it.

Out of all the examples of 'behavioural regeneration', the homology with morphogenetic regeneration is shown most clearly in 'morphogenetic behaviour', concerned with the making of characteristic structures. In some cases animals mend these structures after they have been damaged. For example, potter wasps have been observed to fill in holes made by the experimenter in the walls of their pots, sometimes by means of actions that are never normally performed when the pots are being constructed.[23] And termites repair damage to their galleries and nests through the co-operative and co-ordinated activities of many individual insects.[24]

Activities such as these have sometimes been interpreted as evidence of intelligence, on the ground that animals behaving in a rigidly fixed instinctive manner would not be able to respond so flexibly to unusual situations.[25] But by the same token, regulating sea-urchin embryos and regenerating flatworms could also be said to exhibit intelligence. However, this extension of psychological terminology would be more confusing than helpful. From the point of view of the hypothesis of formative causation, the similarities can be recognized but interpreted the other way round. Seen against the background of morphogenetic regulation and regeneration, the ability of animals to reach behavioural goals in unusual ways raises no fundamentally new principles. And when, in higher animals, certain types of behaviour no longer follow standard chreodes – when behavioural regulation becomes, as it were, the rule rather than the exception – this flexibility can be seen as an extension of the possibilities inherent in morphogenetic and motor fields by their very nature.

Notes

1 For the classical account, see Darwin (1880).
2 Darwin (1882).
3 Audus (1979).
4 Curry (1968).
5 Jaffé (1973).
6 Siegelman (1968).
7 Bünning (1973).

8 Satter (1979).
9 Bose (1926); Roblin (1979).
10 Bentrup (1979).
11 Different species of *Amoeba* differ in detail in their pattern of movement and response from the well-known *A. proteus* type; thus *A. limax* forms few pseudopodia and usually moves forward as a single elongated mass; *A. verrucosa* moves slowly with an almost constant form; *A. velata* generally sends out a free feeler-like pseudopodium into the water. Nevertheless, the general principles of movement appear to be the same. For further details and references, see Jennings (1906).
12 F.D. Warner in Roberts and Hyams (eds) (1979).
13 Sleigh (1968).
14 Jennings (1906).
15 Eckert (1972).
16 E.g. Pecher (1939).
17 Verveen and de Felice (1974).
18 Katz and Miledi (1970).
19 Stevens (1977).
20 Katz (1966).
21 Lindauer (1961).
22 Jennings (1906).
23 Hingston (1928).
24 Marais (1971); von Frisch (1975).
25 Hingston (1928).

10 Instinct and Learning

10.1 The influence of past actions

Motor fields, like morphogenetic fields, are given by morphic resonance from previous similar systems. The detailed structure of an animal and the patterns of oscillatory activity within its nervous system will generally resemble *itself* more closely than any other animal. Thus the most specific morphic resonance acting upon it will be that from its own past (cf. Section 6.5). The next most specific resonance will come from genetically similar animals which lived in the same environment, and the least specific from animals of other races living in different environments. In the valley model of the chreode (Fig. 5), the latter will stabilize the general outline of the valley, while the more specific resonance will determine the detailed topology of the valley bottom.

The 'contours' of the chreodic valley depend on the degree of similarity between the behaviour of similar animals of the same race or species. If their patterns of movement show little variation, morphic resonance will give rise to deep and narrow chreodes, represented by steep-walled valleys (Fig. 27 A). These will have a strongly canalizing effect on the behaviour of subsequent individuals, which will therefore tend to behave in very similar ways. Stereotyped patterns of movement brought about by such chreodes at lower levels appear as reflexes, and at higher levels as instincts.

On the other hand, if similar animals reach the final forms of their motor fields by different patterns of movement, the chreodes will not be so well defined (Fig. 27 B); there will therefore be more scope for individual differences in behaviour. But once a particular animal has reached the behavioural goal in its own way, its subsequent behaviour will tend to be canalized in the same way by

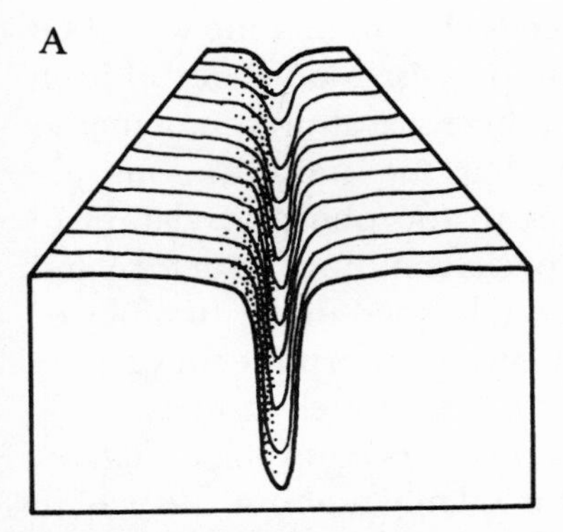

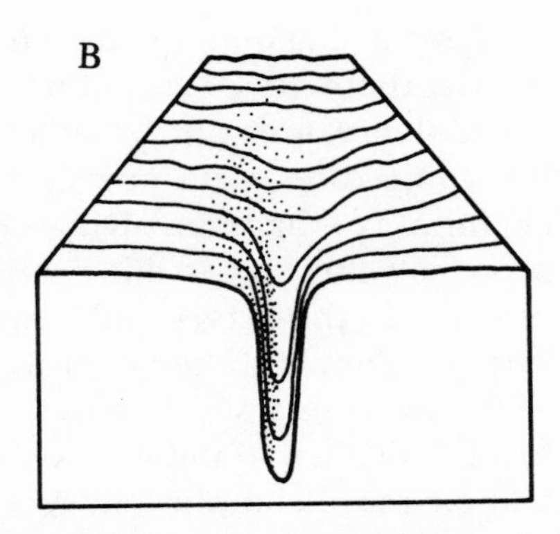

Figure 27 Diagrammatic representation of a deeply canalized chreode (A), and a chreode which is weakly canalized in the initial stages (B).

morphic resonance from its own past states; and the more often these actions are repeated, the stronger will this canalization become. Such characteristic individual chreodes reveal themselves as habits.

Thus, from the point of view of the hypothesis of formative causation, there is only a difference of degree between instincts and habits: both depend on morphic resonance, the former from countless previous individuals of the same species, and the latter mainly from past states of the same individual.

This is not to say that reflexes and instincts do not depend on a very specifically patterned morphogenesis of the nervous system. Obviously they do. Nor is it to say that during processes of learning no physical or chemical changes occur in the nervous system which facilitate the repetition of a pattern of movement. Perhaps in simple nervous systems carrying out stereotyped functions, the potential for such changes may already be 'built in' to the 'wiring' in such a way that learning occurs quasi-mechanistically. For example in the snail *Aplysia* the structure of the nervous system has been found to be almost identical in different individuals, down to the fine details of the arrangement of excitatory and inhibitory synapses on particular cells. Very simple types of learning occur in connection with the reflex withdrawal of the gill into the mantle cavity, namely habituation to harmless stimuli, and sensitization to harmful ones; as they do so, the activities of particular excitatory and inhibitory synapses acting on individual nerve cells alter in definite ways.[1] Of course, the mere description of these processes does not in itself reveal the reasons for the alterations; these are at present a matter for conjecture. One suggestion is that the modifications are chemical, perhaps involving changes in the phosphorylation of proteins.[2] But how is this detailed specialization of structure and function in the nerves and synapses established in the first place? The problem is shifted over to the realm of morphogenesis.

The nervous systems of higher animals are much more variable from individual to individual than in invertebrates such as *Aplysia*, and far more complicated. Very little is known about the way in which learned patterns of behaviour are retained,[3] but enough has been found out to make it clear that there can be no simple explanation in terms of specifically localized physical or chemical 'traces' within the nervous tissue.

Numerous investigations have shown that in mammals learned habits often persist after extensive damage to the cerebral cortex or to sub-cortical regions of the brain. Moreover, when loss of memory does occur, it is not closely related to the location of the lesions, but rather to the total amount of tissue destroyed. K.S. Lashley summed up the results of hundreds of experiments as follows:

> 'It is not possible to demonstrate the isolated localization of a memory trace anywhere within the nervous system. Limited regions may be essential for learning or retention of a particular activity, but within such regions the parts are functionally equivalent.'[4]

A similar phenomenon has been demonstrated in an invertebrate, the octopus: observations on the survival of learned habits after destruction of various parts of the vertical lobe of the brain have led to the seemingly paradoxical conclusion that 'memory is both everywhere and nowhere in particular'.[5]

These findings are extremely puzzling from a mechanistic point of view. In an attempt to account for them, it has been suggested that memory 'traces' are somehow distributed within the brain in a manner analogous to the storage of information in the form of interference patterns in a hologram.[6] But this remains no more than a vague speculation.

The hypothesis of formative causation provides an alternative interpretation, in the light of which the persistence of learned habits in spite of damage to the brain is far less puzzling: the habits depend on motor fields which are not stored within the brain at all, but are given directly from its past states by morphic resonance.

Some of the implications of the hypothesis of formative causation in relation to instinct and learning are considered in the following sections; and in Chapter 11 ways are suggested in which predictions deduced from this hypothesis could be distinguished from those of the mechanistic theory by experiment.

10.2 Instinct

In all animals certain patterns of motor activity are inborn rather than learned. The most fundamental are those of the internal organs, such as the heart and gut, but many of the patterns of movement of limbs, wings and other motor structures are also innate. This is most clearly apparent when animals are able to move around competently almost as soon as they are born or hatched.

It is not always easy to make the distinction between inborn and learned behaviour. In general, characteristic behaviour which develops in young animals reared in isolation can usually be regarded as innate; on the other hand, behaviour which appears only when they are in contact with other members of their species may also be innate, but require stimuli from the other animals for its expression.

Studies of the instinctive behaviour of a wide range of animals have led to several general conclusions, which constitute the classical concepts of ethology.[7] These can be summarized as follows:

(i) Instincts are organized in a hierarchy of 'systems' or 'centres' superimposed upon one another. Each level is primarily activated by a system at the level above it. The highest centre of each of the major instincts may be influenced by a number of factors including hormones, sensory stimuli from the viscera of the animal, and stimuli from the environment.

(ii) The behaviour which occurs under the influence of the major instincts often consists of chains of more or less stereotyped patterns of behaviour called *fixed action patterns*. When such a fixed action pattern constitutes the end-point of a major or minor chain of instinctive behaviour it is called a *consummatory act*. The behaviour in the earlier part of an instinctive chain of behaviour, e.g. searching for food, may be more flexible, and is usually called *appetitive behaviour*.

(iii) Each system requires a specific stimulus in order to be activated or *released*. This stimulus or releaser may come from within the animal's body, or from the environment. In the latter case it is often referred to as a *sign stimulus*. A given releaser or sign stimulus is presumed to act on a specific neuro-sensory mechanism called the *innate releasing mechanism*, which releases the reaction.

These concepts harmonize remarkably well with the ideas of motor fields developed in the previous chapter. The fixed action patterns can be understood in terms of chreodes, and the innate releasing mechanisms as the germ structures of the appropriate motor fields.

10.3 Sign stimuli

The instinctive responses of animals to sign stimuli show that they somehow abstract certain specific and repeatable features from their environments:

> 'An animal responds "blindly" to only part of the total environmental situation and neglects other parts, although its sense organs are perfectly able to receive them . . . These effective stimuli can easily be found by testing the response to various situations differing in one or other of the possible stimuli. Moreover, even when a sense organ is involved in releasing a reaction, only part of the stimuli that it can receive are actually effective. As a rule, an instinctive reaction responds to only very few stimuli, and the greater part of the environment has little or no influence, even though the animal may have the sensory equipment for receiving numerous details.' (N. Tinbergen.[8])

The following examples[9] illustrate these principles:

The aggressive reaction of male stickleback fish during the

breeding season to other male sticklebacks is released mainly by the sign stimulus of the red belly: models with very crude shapes but with red bellies are attacked much more than models with the correct shape but no red colouration. Similar results have been obtained in experiments with the red-breasted robin: a territory-holding male threatens very approximate models with red breasts, or even a mere bundle of red feathers, but responds much less to accurate models without red breasts.

Young ducks and geese react instinctively to the approach of birds of prey, in a manner that depends on the shape of the bird in flight. Studies with cardboard models have shown that the most important feature is a short neck – characteristic of hawks and other predatory birds – while the size and shape of the wings and tail are relatively unimportant.

In certain moths, the sex odour or pheromone normally produced by females causes males to attempt to copulate with any object bearing it.

In locusts of the species *Ephippiger ephippiger*, males attract females which are willing to mate by their song. Females are attracted to singing males from a considerable distance, but ignore silent males even when quite near. Males that are silenced by glueing their wings together are incapable of attracting females.

Hens come to the rescue of chicks in response to their distress call, but not if they simply see them in distress, for example behind a soundproof glass barrier.

According the hypothesis of formative causation, recognition of these sign stimuli must depend on morphic resonance from previous similar animals exposed to similar stimuli. Owing to the process of automatic averaging, this resonance will emphasize only the common features of the spatio-temporal patterns of activity brought about by these stimuli in the nervous system. The result will be that only certain specific stimuli are abstracted from the environment, while others are ignored. Consider, for example, the stimuli acting on hens whose chicks are in distress. Imagine a collection of photographs taken of chicks in distress on many different occasions. Those taken at night will show nothing; those in the daytime will show chicks of different sizes and shapes seen from the front, the rear, the side, or from above; moreover they may be near to other objects of all shapes and sizes, or even concealed behind them. Now if the negatives of all these photographs are superimposed to produce a composite image, no features whatever will be reinforced; the result will simply be a blur. By contrast, imagine a series of tape-recordings made at the same time the photographs were taken. All bear the record of distress calls, and if these sounds are superimposed they reinforce each other to give an automatically averaged distress call. This superimposition of photographs and tape-recordings is analogous to the effects of morphic resonance from the nervous systems of previous hens on a subsequent hen exposed to stimuli from a chick in distress: the visual stimuli result in no specific resonance and evoke no instinctive reaction, however pathetic the chick may look to a human observer, whereas the auditory stimuli do.

This example serves to illustrate what seems to be a general principle: *shapes* are very often ineffective as sign stimuli. The probable reason is that they are highly variable because they depend on the angle from which things are seen. By contrast, colours are much less critically dependent on viewpoint, and sounds and odours hardly at all. Significantly, colours, sounds and odours play important roles as releasers of instinctive reactions; and in those cases where shapes *are* effective, there is a certain constancy of view-point. For example, young birds on the ground see predators flying above them in silhouette, and do indeed respond to such shapes. And when shapes or patterns serve as sexual sign stimuli, they do so in courtship displays or 'presentations' in which animals take up definite stances or postures in relation to their potential mates. The same is true of displays of submission and aggression.

10.4 Learning

Learning can be said to occur when there is any relatively permanent adaptive change in behaviour as a result of past experience. Four general categories can be distinguished:[10]

(i) The most universal type, found even in unicellular organisms,[11] is habituation, which can be defined as the waning of a response as a result of repeated stimulation which is not followed by any kind of reinforcement. A common example is the fading of alarm or avoidance responses to new stimuli which turn out to be harmless: animals get used to them.

This phenomenon implies the existence of some sort of memory, which enables the stimuli to be recognized when they recur. On the hypothesis of formative causation this recognition is primarily due to the morphic resonance of the organism with its own past states, including those brought about by new sensory stimuli. This resonance serves to maintain, and indeed define, the identity of the organism with itself in the past (Section 6.5). Repeated stimuli from the environment to which responses are not reinforced will effectively become part of the organism's own 'background'. Conversely, any new features of the environment will stand out because they are not so recognized: usually the animal will respond with avoidance or alarm precisely because the stimuli are unfamiliar.

In the case of certain stereotyped responses, such as the withdrawal reflex of the gill in the snail *Aplysia*, habituation may occur in a quasi-mechanistic manner on the basis of pre-existing structural and biochemical specializations in the nervous system (Section 10.1). But if so, this specialization is secondary, and seems likely to have evolved from a situation in which habituation depended more directly on morphic resonance.

(ii) In all animals, innate patterns of motor activity appear as the individuals grow up. While some are carried out perfectly the first time they are performed, others improve with time. A young bird's first attempts to fly, for example, or a young mammal's first attempts to walk may be only partially successful, but they get better after repeated attempts. Not all such improvement is due to practice: in some cases it is simply a matter of maturation and occurs just as much with the passage of time in animals which have been immobilized.[12] Nevertheless, many types of motor skill do improve in a way that cannot be attributed to maturation.

From the point of view of the hypothesis of formative causation, this type of learning can be interpreted in terms of behavioural regulation. Morphic resonance from countless past members of the species gives an automatically averaged chreode, which governs an animal's first attempts to carry out a particular innate pattern of movement. This standard chreode may give only approximately satisfactory results, for example because of deviations from the norm in the animal's sense organs, nervous system or motor structures. As the movements are performed, regulation will spontaneously bring about 'fine adjustments' to the overall chreode, and to the lower-level chreodes under its control. These 'adjusted' chreodes will be stabilized by morphic resonance with the animal's own past states as the pattern of behaviour is repeated.

(iii) Animals may come to respond to a stimulus with a reaction which is normally evoked by a different stimulus. This type of learning occurs when the new stimulus is applied at the same time as, or immediately before, the original one. The classical examples are the 'conditioned reflexes' established by I.P. Pavlov in dogs. For instance, the dogs salivated when they were presentedwith food. On repeated occasions a bell was rung as the food was presented, and after some time they began to salivate at the sound of the bell even in the absence of food.

An extreme example of this type of learning occurs in the 'imprinting' of young birds, especially ducklings and goslings. Soon after hatching, they respond instinctively to any reasonably large moving object by following it. Normally this is their mother; but they will also follow foster mothers, human beings, or even inanimate objects which are dragged in front of them. After a relatively short time, they come to recognize the general features of the moving object, and later the specific features. The response of following is then elicited only by the particular bird, person or object with which they have become imprinted.

Analogously, animals often learn to recognize the individual features of their mates or their young by sight, sound, smell or touch. This recognition takes time to develop: for example, a pair of coots with newly-hatched chicks will feed and even adopt strange chicks similar in appearance to their own; but when their young are about two weeks old, they recognize them individually, and henceforth tolerate no strangers, however similar.[13]

A comparable process is probably responsible for the recognition of particular places, such as nest sites, by means of landmarks and other features associated with them. Indeed this type of learning seems likely to play an important part in the development of visual recognition in general. Since the stimuli from an object differ according to the angle from which it is viewed, the animal must learn that they are all connected with the same thing. Likewise, the associations between different kinds of sensory stimulus from the same object – visual, auditory, olfactory, gustatory and tactile – usually have to be learned.

When the new stimulus and the original stimulus occur simultaneously, it might at first sight seem likely that the different patterns of physico-chemical change they bring about in the brain gradually become linked with each other as a result of frequent repetition. But two difficulties stand in the way of this apparently simple interpretation. First, the new stimulus might not be simultaneous with the usual one, but precede it. In this case, it seems necessary to suppose that the influence of the stimulus persists for a while, so that it is still present when the usual stimulus occurs. This kind of memory can be thought of by analogy with an echo that gradually dies away. The existence of such a short-term memory has in fact been demonstrated empirically;[14] it could conceivably be explicable in terms of reverberating circuits of electrical activity within the brain.[15]

Second, associative learning seems to involve definite discontinuities: it occurs in steps, or stages. This may be because the linkage between the new and the original stimulus involves the establishment of a new motor field: the field responsible for the original response must somehow be enlarged to incorporate the new stimulus. In effect, a *synthesis* occurs in which a new motor unit comes into being. And a new unit cannot emerge gradually, but only by a sudden 'quantum jump' (or by several successive 'jumps').

(iv) As well as learning to respond to a particular stimulus *after* they have received it, animals may also learn to behave in such a way that they reach a goal as a *result* of their activities. In the language of the Behaviourist school, in this 'operant conditioning' the response 'emitted' by the animal precedes the reinforcing stimulus. The classical examples are provided by rats in 'Skinner boxes'. These boxes contain a lever which, when pressed, releases a pellet of food. After repeated trials, rats learn to associate the pressing of the lever with the reward. Similarly, they can learn to press a lever in order to avoid the painful stimulus of an electric shock.

The association of a particular pattern of movement with a

reward or with the avoidance of punishment usually seems to happen as a result of trial and error. But intelligence of an altogether higher order has been demonstrated in primates, especially chimpanzees. In some well-known experiments conducted over fifty years ago, W. Köhler found that these apes were capable of solving problems in an 'insightful' way.[16] For example, chimpanzees were placed in a high chamber with unclimbable walls, from the ceiling of which hung a bunch of ripe bananas, too high for them to reach. After a number of attempts to get the fruit by standing on their hind legs and by jumping, they gave up this approach. After a while, one or other of the apes would glance first at one of a number of wooden boxes which had been placed in the chamber at the beginning of the experiment, and then at the bananas. He would then drag the box underneath them and stand on it. This did not bring him high enough, so he fetched another box and put it on top of the first, but it was still not high enough; he then added a third, climbed up, and grabbed the fruit.

Many more examples of insightful behaviour have been demonstrated by subsequent investigators: in one experiment, for instance, chimpanzees learned to use sticks to rake in food placed outside the cages beyond their reach. They did this sooner if they had been allowed to play with the sticks for several days beforehand; during this period they came to use the sticks as functional extensions of their arms. Thus the use of the sticks to rake in the food represented 'the integration of motor components acquired during earlier experience into new and appropriate behaviour patterns'.[17]

In both 'trial and error' and 'insight' learning, existing chreodes are integrated within new higher-level motor fields. These syntheses can only come about by sudden 'jumps'. If the new patterns of behaviour are successful, they will tend to be repeated. Hence the new motor fields will be stabilized by morphic resonance as the learned behaviour becomes habitual.

10.5 Innate tendencies to learn

The originality of learning may be absolute: a new motor field may come into being not only for the first time in the history of an individual, but for the first time ever. On the other hand, an animal may learn something that other members of its species have already learned in the past. In this case, the emergence of the appropriate motor field may well be facilitated by morphic resonance from previous similar animals. If a motor field becomes increasingly well-established through repetition in many individuals, learning is likely to become progressively easier: there will be a strong innate disposition towards acquiring this particular pattern of behaviour.

Thus learned behaviour which is repeated very frequently will tend to become semi-instinctive. By a converse process, instinctive behaviour may come to be semi-learned. This latter type of intergradation between instinctive and learned behaviour is illustrated particularly clearly by the songs of birds.[18] In some species, such as the wood pigeon and the cuckoo, the pattern of the song is almost completely innate, and hence varies little from bird to bird. But in others, for example the chaffinch, while the song has a general structure characteristic of the species, in its fine detail it differs from individual to individual; these differences can be recognized by other birds and play an important part in the birds' family and social life. Birds raised in isolation produce simplified and rather featureless versions of the song of their species, showing that its general structure is innate. However, under normal conditions they develop and improve their singing by imitating other birds of their own kind. This process is taken much further in mocking birds, for example, which borrow elements from the songs of species other than their own. And some kinds of birds, notably parrots and mynahs, under the artificial conditions of captivity often go so far as to imitate their human foster parents.

In species whose songs are almost entirely innate, the lack of individual variation is both an effect and a cause of the well-defined and highly stabilized motor chreodes (cf. Fig. 27A): the more the same pattern of movement is repeated, the greater will be its tendency to be repeated in the future. But in species with individual differences in song, morphic resonance will give less well-defined chreodes (cf. Fig. 27B): the general structure of the chreode will be given by the process of automatic averaging, but the details will depend on the individual. The pattern of movements it makes when it first begins to sing will influence its subsequent singing, owing to the specificity of morphic resonance from its own past states; with repetition, its characteristic pattern of song will become habitual as its individual chreodes are deepened and stabilized.

Notes

1 Kandel (1979).
2 ibid.
3 H.A. Buchtel and G. Berlucchi in Duncan and Weston-Smith (eds) (1977).
4 Lashley (1950), p.478.
5 Boycott (1965).
6 Pribram (1971).
7 For a comprehensive review and discussion, see Thorpe (1963).
8 Tinbergen (1951), p.27.
9 ibid.
10 Thorpe (1963).
11 E.g. Jennings (1906).
12 Hinde (1966).
13 Thorpe (1963), p.429.
14 Spear (1978).
15 Although this idea, suggested by Hebb (1949), has been advocated for many years, it has neither been conclusively refuted nor convincingly supported by experimental evidence.
16 Köhler (1925).
17 Loizos (1967), p.203.
18 Thorpe (1963).

11 The Inheritance and Evolution of Behaviour

11.1 The inheritance of behaviour

On the hypothesis of formative causation, the inheritance of behaviour depends on genetic inheritance, *and* on the morphogenetic fields which control the development of the nervous system and the animal as a whole, *and* on the motor fields given by morphic resonance from previous similar animals. By contrast, according to the conventional theory innate behaviour is supposed to be 'programmed' in the DNA.

Relatively few experimental investigations have been carried out on the inheritance of behaviour, largely because it is difficult to quantify. Nevertheless, various attempts have been made: for instance, in experiments with rats and mice behaviour has been 'measured' in terms of their running speed in treadmills; the frequency and duration of sexual activity; defecation scores, defined as the number of faecal boluses deposited in a given area in unit time; maze learning abilities; and susceptibility to audiogenic seizures, caused by very loud noises. A heritable component of these responses has been demonstrated by breeding from animals with high or low scores: the progeny tend to have scores resembling those of their parents.[1] The trouble with investigations of this type is that they reveal very little about the inheritance of *patterns* of behaviour; moreover, the results are difficult to interpret because they are open to influence by so many different factors. For example, a lower treadmill speed or a reduced frequency of mating could simply be due to a general reduction in vigour as a consequence of a heritable metabolic deficiency.

In some cases, the reasons for genetic alterations of behaviour have been investigated in considerable detail. In the small nematode worm *Caenorhabditis*, certain mutants that wriggle abnormally show structural changes in their nervous systems.[2] In *Drosophila*, various 'behavioural mutations' which abolish the normal response to light have been found to affect the photo-receptors or the peripheral visual neurons.[3] In mice, a number of mutations are known to affect the morphogenesis of the nervous system, leading to defects of whole regions of the brain. In human beings, various congenital abnormalities of the nervous system are associated with abnormal behaviour, for example in Down's syndrome, a type of mongolism. And then behaviour can also be affected by hereditary physiological and biochemical defects; for instance, in man the condition of phenylketonuria, associated with mental retardation, is due to a deficiency of the enzyme phenylalanine hydroxylase.

The fact that innate behaviour is affected by genetically determined alterations in the structure and function of the sense organs, nervous system, etc., does not, of course, prove that its inheritance is explicable in terms of genetic factors alone; it only shows that a normal body is necessary for normal behaviour. Think again of the radio analogy: changes within the set affect its performance; but this does not prove that the music which comes out of the loudspeakers originates inside the set itself.

In the realm of behaviour, biochemical, physiological and anatomical changes may prevent the appearance of germ structures, and hence whole motor fields may fail to act; or they may have various quantitative effects on the movements controlled by these fields. And, in fact, investigations on the inheritance of fixed action patterns show that 'it is not difficult to find variations which affect the performance in a minor fashion, but the unit still appears in a clearly recognizable form if it appears at all'.[4]

The inheritance of motor fields is probably dependent on the factors already discussed in connection with the inheritance of morphogenetic fields (Chapter 7). Generally speaking, in hybrids between two races or species, the dominance of the motor fields of one over those of the other is likely to depend on the relative strength of the morphic resonance from the parental types (cf. Fig. 19). If one belongs to a well-established race or species, and the other to a relatively new one with a small past population, the motor fields of the former would be expected to be dominant. But if the parental races or species are equally well-established, the hybrids would be expected to come under the influence of both to a similar extent.

This is in fact what seems to happen. In some cases the results are quite bizarre, because the patterns of behaviour of the parental types are incompatible with each other. One example is provided by the hybrids produced by crossing two kinds of lovebird. Both parental species make their nests out of strips which they tear from leaves in a similar manner, but whereas one (Fischer's lovebird) then carries these strips to the nest in its bill, the other (the peach-faced lovebird) carries them tucked in among its feathers. Hybrids tear the strips from the leaves normally, but then behave in a most confused manner, sometimes tucking the strips in among their feathers, sometimes carrying them in their bills; but even when they carry them in their bills, they erect the feathers of the lower back and rump and attempt to tuck them in.[5]

11.2 Morphic resonance and behaviour: an experimental test

In mechanistic biology, a sharp distinction is drawn between innate and learned behaviour: the former is assumed to be 'genetically programmed' or 'coded' in the DNA, while the latter is supposed to result from physico-chemical changes in the nervous system. There is no conceivable way in which such changes could specifically modify the DNA (as the Lamarckian theory would require); it is therefore considered impossible for the learned behaviour acquired by an animal to be inherited by its offspring (excluding, of course, 'cultural inheritance', whereby the offspring learn patterns of behaviour from their parents).

By contrast, according to the hypothesis of formative causation, there is no difference in kind between innate and learned behaviour, in that both depend on motor fields given by morphic resonance (Section 10.1). This hypothesis therefore admits a possible transmission of learned behaviour from one animal to another, and leads to testable predictions which differ not only from those of the orthodox theory of inheritance, but also from those of the Lamarckian theory.

Consider the following experiment. Animals of an inbred strain are placed under conditions in which they learn to respond to a given stimulus in a characteristic way. They are then made to repeat this pattern of behaviour many times. *Ex hypothesi*, the new motor field will be reinforced by morphic resonance, which will not only cause the behaviour of the trained animals to become increasingly habitual, but will also affect, although less specifically, any similar animal exposed to a similar stimulus: the larger the number of animals in the past which have learned the task, the easier it should be for subsequent similar animals to learn it. Therefore in an experiment of this type it should be possible to observe a progressive increase in the rate of learning not only in animals descended from trained ancestors, but also in genetically similar animals descended from untrained ancestors. This prediction differs from that of the Lamarckian theory, according to which only the descendants of trained animals should learn quicker. And on the conventional theory, there should be no increase in the rate of learning of the descendants of untrained *or* trained animals.

To summarize: an increased rate of learning in successive generations of both trained and untrained lines would support the hypothesis of formative causation; an increase only in trained lines, the Lamarckian theory; and an increase in neither, the orthodox theory.

Experiments of this type have in fact already been performed. The results support the hypothesis of formative causation.

The original experiment was started by W. McDougall at Harvard in 1920, in the hope of providing a thorough test of the possibility of Lamarckian inheritance. The experimental animals were white rats, of the Wistar strain, which had been carefully inbred under laboratory conditions for many generations. Their task was to learn to escape from a specially constructed tank of water by swimming to one of two gangways which led out of the water. The 'wrong' gangway was brightly illuminated, while the 'right' gangway was not. If the rat left by the illuminated gangway it received an electric shock. The two gangways were illuminated alternately, one on one occasion, the other on the next. The number of errors made by a rat before it learned to leave the tank by the non-illuminated gangway gave a measure of its rate of learning:

> 'Some of the rats required as many as 330 immersions, involving approximately half that number of shocks, before they learnt to avoid the bright gangway. The process of learning was in all cases one which suddenly reached a critical

> point. For a long time the animal would show clear evidence of aversion for the bright gangway, frequently hesitating before it, turning back from it, or taking it with a desperate rush; but, not having grasped the simple relation of constant correlation between bright light and shock, he would continue to take the bright route as often or nearly as often as the other. Then, at last, would come a point in his training at which he would, if he found himself facing the bright light, definitely and decisively turn about, seek the other passage, and quietly climb out by the dim gangway. After attaining this point, no animal made the error of again taking the bright gangway, or only in very rare instances.'[6]

In each generation, the rats from which the next generation were to be bred were selected at random *before* their rate of learning was measured, although mating took place only after they were tested. This procedure was adopted to avoid any possibility of conscious or unconscious selection in favour of quicker-learning rats.

This experiment was continued for 32 generations and took 15 years to complete. In accordance with the Lamarckian theory, there was a marked tendency for rats in successive generations to learn more quickly. This is indicated by the average number of errors made by rats in the first eight generations, which was over 56, compared with 41, 29 and 20 in the second, third and fourth groups of eight generations, respectively.[7] The difference was apparent not only in the quantitative results, but also in the actual behaviour of the rats, which became more cautious and tentative in the later generations.[8]

McDougall anticipated the criticism that in spite of his random selection of parents in each generation, some sort of selection in favour of quicker-learning rats could nevertheless have crept in. In order to test this possibility, he started a new experiment, with a different batch of rats, in which parents were indeed selected on the basis of their learning score. In one series, only quick learners were bred from in each generation, and in the other series only slow learners. As expected, the progeny of the quick learners tended to learn relatively quickly, while the progeny of the slow learners learned relatively slowly. However, even in the latter series, the performance of the later generations improved very markedly, in spite of repeated selection in favour of slow learning (Fig. 28).

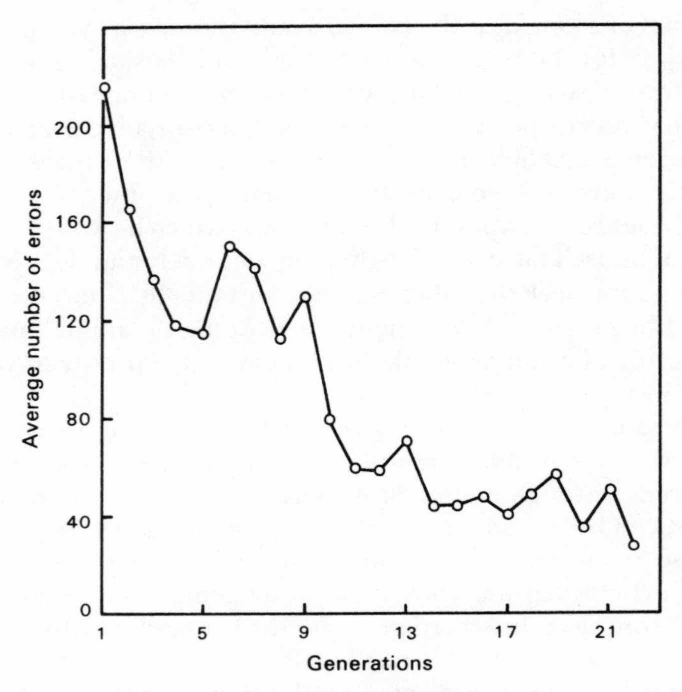

Figure 28 The average number of errors in successive generations of rats selected in each generation for slowness of learning. (Data from McDougall, 1938).

These experiments were done carefully, and critics were unable to dismiss the results on the ground of flaws in technique. But they did draw attention to a weakness in the experimental design: McDougall had failed to test systematically the change in the rate of learning of rats whose parents had not been trained.

One of these critics, F.A.E. Crew, of Edinburgh, repeated McDougall's experiment with rats derived from the same inbred strain, using a tank of similar design. He included a parallel line of 'untrained' rats, some of which were tested in each generation for their rate of learning, while others, which were not tested, served as the parents of the next. Over the 18 generations of this experiment, Crew found no systematic change in the rate of learning either in the trained or in the untrained line.[9] At first, this seemed to cast serious doubt on McDougall's findings. However, Crew's results were not directly comparable in three important respects. First, for some reason the rats found it much easier to learn the task in his experiment than in the earlier generations of McDougall's. So pronounced was this effect that a considerable number of rats in both trained and untrained lines 'learned' the task immediately without receiving a single shock! The average scores of Crew's rats right from the beginning were similar to those of McDougall's after more than 30 generations of training. Neither Crew nor McDougall was able to provide a satisfactory explanation of this discrepancy. But, as McDougall pointed out, since the purpose of the investigation was to bring to light any effect of training on subsequent generations, an experiment in which some rats received no training at all and many others received very little would not be qualified to demonstrate this effect.[10] Second, Crew's results showed large and apparently random fluctuations from generation to generation, far larger than the fluctuations in McDougall's results, which could well have obscured any tendency to improve in the scores of later generations. Third, Crew adopted a policy of very intensive inbreeding, crossing only brothers with their sisters in each generation. He had not expected this to have adverse effects, since the rats came from an inbred stock to start with:

> 'Yet the history of my stock reads like an experiment in inbreeding. There is a broad base of family lines and a narrow apex of two remaining lines. The reproductive rate falls and line after line becomes extinct.'[11]

Even in the surviving lines, a considerable number of animals were born with such extreme abnormalities that they had to be discarded. The harmful effects of this severe inbreeding could well have masked any tendency for the rate of learning to improve. Altogether, these defects in Crew's experiment mean that the results can only be regarded as inconclusive; and in fact he himself was of the opinion that the question remained open.[12]

Fortunately, this is not the end of the story. The experiment was carried out again by W.E. Agar and his colleagues at Melbourne, using methods which did not suffer from the disadvantages of Crew's. Over a period of 20 years, they measured the rates of learning of trained and untrained lines for 50 successive generations. In agreement with McDougall, they found that there was a marked tendency for rats of the trained line to learn more quickly in subsequent generations. *But exactly the same tendency was also found in the untrained line.*[13]

It might be wondered why McDougall did not also observe a similar effect in his own untrained lines. The answer is that he did. Although he tested control rats from the original untrained stock only occasionally, he noticed 'the disturbing fact that the groups of controls derived from this stock in the years 1926, 1927, 1930 and 1932 show a diminution in the average number of errors from 1927 to 1932'. He thought this result was probably fortuitous, but added:

> 'It is just possible that the falling off in the average number of errors from 1927 to 1932 represents a real change of constitution of the whole stock, an improvement of it (with respect to this particular faculty) whose nature I am unable to suggest.'[14]

With the publication of the final report by Agar's group in 1954 the prolonged controversy over 'McDougall's Lamarckian Experiment' came to an end. The similar improvement in both trained and untrained lines ruled out a Lamarckian interpretation. McDougall's *conclusion* was refuted, and that seemed to be the end of the matter. On the other hand, his *results* were confirmed.

These results seemed completely inexplicable; they made no sense in terms of any current ideas, and they were never followed up. But they make very good sense in the light of the hypothesis of formative causation. Of course they cannot in themselves *prove* the hypothesis; it is always possible to suggest other explanations, for example that the successive generations of rats became increasingly intelligent for an unknown reason unconnected with their training.[15]

In future experiments, the most unambiguous way of testing for the effects of morphic resonance would probably be to cause large numbers of rats (or any other animals) to learn a new task in one location; and then see if there was an increase in the rate at which similar rats learned to carry out the same task at another location hundreds of miles away. The initial rate of learning at both locations should be more or less the same. Then, according to the hypothesis of formative causation, the rate of learning should increase progressively at the location when large numbers are trained; and a similar increase should also be detectable in the rats at the second location, even though very few rats had been trained there. Obviously, precautions would need to be taken to avoid any possible conscious or unconscious bias on the part of the experimenters. One way would be for experimenters at the second location to test the rate of learning of rats in several *different* tasks, at regular intervals, say monthly. Then at the first location, the particular task in which thousands of rats would be trained would be chosen at random from this set. Moreover, the time at which the training began would also be selected at random; it might, for example, be four months after the regular tests began at the second location. The experimenters at the second location would not be told either which task had been selected, nor when the training had begun at the first location. If, under these conditions, a marked increase in the rate of learning in the selected task were detected at the second location after the training had begun at the first, then this result would provide strong evidence in favour of the hypothesis of formative causation.

An effect of this type might well have occurred when Crew and Agar's group repeated McDougall's work. In both cases, their rats started off learning the task considerably more quickly than McDougall's rats did when he first began his experiment.[16]

If the experiment proposed above were actually performed, and if it gave positive results, it would not be fully reproducible by its very nature; for in attempts to repeat it, the rats would be influenced by morphic resonance from the rats in the original experiment. To demonstrate the same effect again and again, it would be necessary to change either the task or the species used in each experiment.

11.3 The evolution of behaviour

Whereas the Fossil Record provides direct evidence about the structure of past animals, it reveals practically nothing about their behaviour. Consequently most ideas about the evolution of behaviour cannot be based on evidence from the past, but only on comparisons between species in existence at present. Thus, for example, theories can be constructed about the evolution of social behaviour in the bees by comparing existing social species with solitary and colonial species, which are presumed to be more primitive. But however reasonable such theories may seem, they can never be more than speculative.[17] Morover, theories of behavioural evolution depend on *assumptions* about the way in which behaviour is inherited, since so little is actually known.

The mechanistic, or neo-Darwinian, theory assumes that innate behaviour is 'programmed' or 'coded' in the DNA, and that new types of behaviour are caused by chance mutations. Then natural selection favours favourable mutants; hence instincts evolve. Chance mutations are also assumed to give animals capacities for particular types of learning. Then animals whose survival and reproduction benefits from these capacities are favoured by natural selection. Hence capacities for learning evolve. Even a tendency for learned behaviour to become innate can be attributed to chance mutations, by the hypothetical Baldwin effect: animals may respond to new situations by learning to behave in appropriate ways; chance mutations which cause this behaviour to appear without the need for learning will be favoured by natural selection; hence behaviour which was at first learned may become innate, not because of an inheritance of acquired characteristics, but because appropriate mutations happened to occur.

There seems to be practically no limit to what can be accounted for by the invocation of favourable chance mutations which change the 'genetic programming' of behaviour. Then these neo-Darwinian theories can be developed in a mathematical form by means of calculations based on the formulae of theoretical population genetics.[18] But in so far as these speculations are untestable, they have no independent value; they merely elaborate the mechanistic assumptions from which they start.

The hypothesis of formative causation leads to very different interpretations of the evolution of behaviour. To the extent that genetical changes influence behaviour, natural selection would still be expected to lead to alterations in the 'gene pools' of populations. But the specific patterns of behaviour themselves depend on the inheritance of motor fields by morphic resonance. The more a given pattern of behaviour is repeated, the stronger will this resonance become. Thus the repetition of instinctive behaviour will tend to fix the instincts more and more. On the other hand, if patterns of behaviour vary from individual to individual, morphic resonance will not produce well-defined chreodes; hence the behaviour will be less stereotyped. And the greater the variety of behaviour, the greater will be the scope for variation in future generations. This type of evolution, in a direction permitting the emergence of intelligence, has taken place to some extent among the birds, more so in the mammals, and most of all in man.

In some cases, behaviour which is semi-learned must have evolved from a background in which it was fully instinctive. One way in which this could have happened is through the hybridization of races with different chreodes, giving rise to composite motor fields with more scope for individual variation.

In other cases, semi-instinctive behaviour could have evolved from behaviour which was originally learned, as a result of frequent repetition. Consider, for example, the behaviour of different breeds of dog. Sheep dogs have been trained and selected over many generations for the ability to round up sheep, retrievers to retrieve, pointers to point, fox hounds to chase foxes, and so on. Dogs often show an innate tendency towards the behaviour characteristic of their breed even before they are trained.[19] Perhaps these tendencies are not quite strong enough to be called instincts, but they are strong enough to show that there is only a difference of degree between instinct and a hereditary pre-disposition to learn particular types of behaviour. Of course, breeds of dogs have evolved under conditions of artificial rather than natural selection, but the same principles seem likely to apply in both cases.

While it is relatively easy to imagine how some types of instinctive behaviour could have developed by the repetition of learned behaviour generation after generation, this cannot feasibly account for the evolution of all types of instinct, especially in animals with a very limited capacity for learning. Possibly some new instincts emerged from new permutations and combinations of pre-existing instincts; one way in which this could occur would be through hybridization between races or species with different patterns of behaviour. Another way in which new combinations might come about is through the incorporation of 'displacement activities', the seemingly irrelevant actions performed by animals 'torn' between conflicting instincts. Certain elements of courtship rituals may well have originated in this way.[20] It is also conceivable that mutations or exposure to unusual environments could enable an animal to 'tune in' to other species' motor chreodes (cf. Sections 8.6).

But in addition to the recombination of existing chreodes, there must be some way in which entirely new motor fields come into being in animals whose behaviour is almost entirely instinctive. New patterns of behaviour could only emerge if the usual repetition of ancestral behaviour was blocked, either by a change in the environment, or by a mutation which altered the normal physiology or morphogenesis of the animal. In most such cases the animal might act in an unco-ordinated and ineffective manner; but occasionally a new motor field might come into being. And whenever a new field appears for the first time, there must be 'jump' which cannot be fully accounted for in terms of preceding energetic or formative causes (Sections 5.1, 8.7).

If the pattern of behaviour due to a new motor field impairs the ability of animals to survive and reproduce, it will not be repeated very often; for animals which persist in this behaviour will be eliminated by natural selection. But if the pattern of behaviour helps the animals which perform it to survive and reproduce, it will tend to be repeated frequently and will therefore be increasingly reinforced by morphic resonance. Thus the motor field will be favoured by natural selection.

11.4 Human behaviour

Higher animals often behave more flexibly than lower animals. However, this flexibility is confined to the early stages of a behavioural sequence, and especially to the initial appetitive phase; the later stages, and in particular the final stage, the consummatory act, are performed in a stereotyped manner as fixed action patterns (Section 10.1).

In terms of the landscape model, a major motor field can be represented by a broad valley, which then narrows down and becomes increasingly steep-walled, finally ending up in a deep canyon (Fig. 27 B). The broad valley corresponds to the appetitive phase, in which many alternative pathways can be followed; these pathways then converge as they are 'funnelled' towards the final highly canalized chreode of the consummatory act.

In human behaviour the ranges of ways in which behavioural goals are reached are far wider than in any other species, but the same principles seem to apply: under the influence of the higher-level motor fields, patterns of action are 'funnelled' towards stereotyped consummatory acts which are generally innate. For example, people obtain their food by all sorts of different methods, either directly by hunting, gathering, fishing, herding or farming, or earn it indirectly by performing various tasks or jobs. Then the food is prepared and cooked in many different ways, and placed in the mouth by a variety of means, for instance by hand, or with chopsticks, or on a spoon. But there is little difference in the way the food is chewed, and the consummatory act of the whole motor field of feeding, swallowing, is similar in all men. Likewise, in the behaviour governed by the motor field of reproduction, methods of courtship and systems of marriage differ widely, but the consummatory act of copulation towards which they lead is more or less stereotyped. In the male, the final fixed action pattern, that of ejaculation, proceeds automatically, and is in fact innate.

Thus the very varied patterns of human behaviour are usually directed towards a limited number of goals given by the motor fields inherited from past members of the species by morphic resonance; in general, these goals are related to the development, maintenance or reproduction of the individual or social group. Even play and exploratory activity not immediately directed towards such goals often help achieve them later on, as they do in other species. For neither play nor 'generalized exploratory appetitive behaviour' in the absence of immediate reward is confined to man: rats, for example, explore their environment and investigate objects even when they are satiated.[21]

However, not all human activity is subordinated to the motor fields which canalize it towards biological or social goals; some is explicitly directed towards transcendent ends. This kind of behaviour is shown in its purest form in the lives of saints. But clearly most of the behaviour of most human beings has no such transcendent direction.

Although the range of variation in human behaviour is very wide when the species as a whole is considered, in any given society the activities of individuals tend to fall into a limited number of standard patterns. People usually repeat characteristically structured activities which have already been performed over and over again by many generations of their predecessors. These include the speaking of a particular language; the motor skills associated with hunting, farming, weaving, tool-making, cooking, and so on; songs and dances; and the types of behaviour specific to particular social roles.

All the patterns of activity characteristic of a given culture can be regarded as chreodes.[22] The more often they are repeated, the more strongly stabilized they will be. But because of the bewildering variety of culture-specific chreodes, each of which could potentially canalize the movements of any human being, morphic resonance cannot by itself lead an individual into one set of chreodes rather than another. So none of these patterns of behaviour expresses itself spontaneously: all have to be learned. An individual is *initiated* into particular patterns of behaviour by other members of the society. Then as the process of learning begins, usually by imitation, the performance of a characteristic pattern of movement brings the individual into morphic resonance with all those who have carried out this pattern of movement in the past. Consequently learning is facilitated as the individual 'tunes in' to specific chreodes.

Processes of initiation are indeed traditionally understood in terms rather similar to these. Individuals are thought to enter into states or modes of existence which precede them and have a sort of trans-personal reality.

The facilitation of learning by morphic resonance would be difficult to demonstrate empirically in the case of long-established patterns of behaviour; but a change in the rate of learning should be more readily detectable with motor patterns of recent origin. Thus, for example, within the present century it should have become progressively easier to learn to ride a bicycle, drive a car, play the piano, or use a typewriter, owing to the cumulative morphic resonance from the large number of people who have already acquired these skills. However, even if reliable quantitative data showed that the rates of learning had in fact increased, the interpretation would be complicated by the probable influence of other factors like improved machine design, better teaching methods, and a higher motivation to learn. But with specially designed experiments in which precautions were taken to hold these other factors constant, it might well be possible to obtain persuasive evidence for the predicted effect.

The hypothesis of formative causation applies to all aspects of human behaviour in which particular patterns of movement are repeated. But it cannot account for the origin of these patterns in the first place. Here, as elsewhere, the problem of creativity lies outside the scope of natural science, and an answer can only be given on metaphysical grounds (cf. Sections 5.1, 8.7 and 11.3).

Notes

1 Parsons (1967).
2 Brenner (1973).
3 Benzer (1973).
4 Manning (1975), p.80.
5 Dilger (1962).
6 McDougall (1927), p.282.
7 McDougall (1938).
8 McDougall (1930).
9 Crew (1936).
10 McDougall (1938).
11 Crew (1936), p.75.
12 Tinbergen (1951), p.201.
13 Agar, Drummond, Tiegs and Gunson (1954).
14 Rhine and McDougall (1933), p.223.
15 A number of possible explanations were suggested at the time these experiments were being carried out; they are discussed in McDougall's papers, to which the interested reader should refer. None of these explanations turned out to be plausible on closer examination.

Agar *et al.* (1954) noticed that fluctuations in the rates of learning were associated with changes, extending over several generations, in the health and vigour of the rats. McDougall had already noted a similar effect. A statistical analysis showed that there was indeed a low but significant (at the 1% level of probability) correlation between vigour (measured in terms of fertility) and learning rates in the 'trained' line, but not in the 'untrained' line. However, if only the first forty generations were considered, the coefficients of correlation were somewhat higher: 0.40 in the 'trained' line, and 0.42 in the 'untrained'. But while this correlation may help to account for the fluctuations in the results, it cannot plausibly explain the overall trend. According to standard statistical theory, the proportion of the variation 'explained' by a correlated variable is given by the square of the correlation coefficient, in this case $(0.4)^2=0.16$. In other words, variations in vigour account for only 16% of the changes in the rate of learning.

16 McDougall estimated that the average number of errors in his first generation was over 165. In Crew's experiment this figure was 24, and in Agar's, 72; see the discussions in Crew (1936), and in Agar, Drummond and Tiegs (1942). If Agar's group had used rats of identical parentage and followed the same procedures as Crew, their initial score might have been expected to be even lower than his. However, owing to the different parentage of their rats, and to differences in their testing procedure, the results are not fully comparable. Nevertheless the

greater facility of learning in these later experiments is suggestive.

17 Brown (1975).

18 Numerous examples of this type of speculation can be found in Wilson (1975) and Dawkins (1976).

19 E.g. Clarke (1980).

20 Tinbergen (1951).

21 Thorpe (1963).

22 Language in particular provides an excellent example of the hierarchical organization of motor fields, and a beginning has already been made by R. Thom in developing a theory of language in terms of chreodes; see his *Structural Stability and Morphogenesis*, Chapter 6.

12 Four Possible Conclusions

12.1 The hypothesis of formative causation

The presentation of the hypothesis of formative causation in the preceding chapters of this book can only be regarded as a preliminary sketch: the hypothesis is capable of being worked out in far greater detail both in the realm of biology and of physics. But until some of its predictions have been tested, there will probably be little incentive to undertake this task: only if persuasive positive results are obtained is the hypothesis likely to seem worth pursuing, at least in its present form. Examples of possible experimental tests have been given in Sections 5.6, 7.4, 7.6, 11.2 and 11.4; and more could be devised.

The hypothesis of formative causation is a testable hypothesis about objectively observable regularities of nature. It cannot provide any answers to the questions posed by the origination of new forms and new patterns of behaviour, or by the fact of subjective experience. Such questions can be answered only by theories of reality more far-reaching than those of natural science, in other words by metaphysical theories.

At present, scientific and metaphysical questions are frequently confused with each other, because of the close connection between the mechanistic theory of life and the metaphysical theory of materialism. The latter would still be defensible if the mechanistic theory were to be superseded within biology by the hypothesis of formative causation, or indeed by any other hypothesis. But it would lose its privileged position; it would have to enter into free competition with other metaphysical theories.

In order to illustrate the important distinction between the realms of science and of metaphysics, in the following Sections four different metaphysical theories are briefly outlined. All four are equally compatible with the hypothesis of formative causation and, from the point of view of natural science, the choice between them can only be left entirely open.

12.2 Modified materialism

Materialism starts from the assumption that only matter is real; hence everything that exists is either matter or entirely dependent upon matter for its existence. However, the concept of matter has no fixed meaning; in the light of modern physics it has already been extended to include physical fields, and material particles have come to be regarded as forms of energy. The philosophy of materialism has had to be modified accordingly.

Morphogenetic fields and motor fields are associated with material systems; they too can be regarded as aspects of matter (Section 3.4). Thus materialism could be further modified to incorporate the idea of formative causation.[1] In the following discussion this new form of the materialist philosophy will be referred to as modified materialism.

Materialism denies *a priori* the existence of any non-material causal agency; the physical world is considered to be causally closed. Hence there can be no such thing as a non-material self which acts upon the body, as there seems to be from a subjective point of view. Rather, conscious experience is either in some sense the same thing as material states of the brain, or it simply runs parallel to these states without affecting them.[2] But whereas in conventional materialism brain states are considered to be determined by a combination of energetic causation and chance events, in modified materialism they would, in addition, be determined by formative causation. Indeed, conscious experience would probably best be thought of as an aspect or epiphenomenon of the motor fields acting on the brain.

The subjective experience of free will cannot, *ex hypothesi,* correspond to the causal influence of a non-material self upon the body. However, it is conceivable that some of the random events within the brain might be subjectively experienced as free choices; but this apparent freedom would be nothing but an aspect or epiphenomenon of the chance activation of one motor field rather than another.

If all conscious experience is simply an accompaniment of, or runs parallel to, the motor fields acting upon the brain, then conscious memory, like the memory of motor habits (cf. Section 10.1), must depend on morphic resonance from past states of the brain. Neither conscious nor unconscious memories would be stored within the brain.

In the context of conventional materialism, the evidence for parapsychological phenomena can only be denied, ignored or explained away, in so far as it appears to be inexplicable in terms of energetic causation. But modified materialism might well permit a more positive attitude. For it is not inconceivable that some of these alleged phenomena might turn out to be compatible with the hypothesis of formative causation: in particular, it might be possible to formulate an explanation of telepathy in terms of morphic resonance,[3] and of psychokinensis in terms of the modification of probabilistic events within objects under the influence of motor fields.[4]

The origin of new forms, new patterns of behaviour and new ideas cannot be explained in terms of pre-existing energetic and formative causes (Sections 5.1, 8.7, 11.3, 11.4). Moreover, materialism denies the existence of any non-material creative agency which could have given rise to them. Hence they have no cause. Their origin must therefore be attributed to chance, and evolution can only be seen in terms of the interplay of chance and physical necessity.

In summary, according to this modified philosophy of materialism, the universe is composed of matter and energy, which are either eternal or of unknown origin, organized into an enormous variety of inorganic and organic forms which all arose by chance, governed by laws which cannot themselves be explained. Conscious experience is either an aspect of or runs parallel to the motor fields acting on the brain. All human creativity, like evolutionary creativity, must ultimately be ascribed to chance. Human beings adopt their beliefs (including the belief in materialism) and carry out their actions as a result of chance events and physical necessities within their brains. Human life has no purpose beyond the satisfaction of biological and social needs; nor has the evolution of life, nor the universe as a whole, any purpose or direction.

12.3 The conscious self

Contrary to the philosophy of materialism, the conscious self can be admitted to have a reality which is not merely derivative from matter. One can accept, rather than deny, that one's own conscious self has the capacity to make free choices. Then, by analogy, other people can also be assumed to be conscious beings with a similar capacity.

This 'common sense' view leads to the conclusion that the conscious self and the body *interact*. But then how does this interaction take place?

In the context of the mechanistic theory of life, the conscious self has to be seen as a sort of 'ghost in the machine'.[5] To materialists this notion seems inherently absurd. And even the defenders of the interactionist position have been unable to specify how the interaction takes place, beyond the vague suggestion that it might somehow depend on a modification of quantum events within the brain.[6]

The hypothesis of formative causation enables this long-standing problem to be seen in a new light. The conscious self can be thought of as interacting not with a machine, but with motor fields. These motor fields are associated with the body and depend on its physico-chemical states. But the self is neither the same as the motor fields, nor does its experience simply parallel the changes brought about within the brain by energetic and formative causation. It 'enters into' the motor fields, but it remains over and above them.

Through these fields, the conscious self is closely connected with the external environment and with the states of the body in perception and in consciously controlled activity. But subjective experience which is not directly concerned with the present environment or with immediate action – for example in dreams, reveries, and discursive thinking – need not necessarily bear any particularly close relationship to the energetic and formative causes acting on the brain.

At first sight, this conclusion might appear to contradict the evidence showing that states of consciousness are often associated with characteristic physiological activities. Dreams, for instance, tend to be accompanied by rapid eye movements and by electrical rhythms of particular frequencies within the brain.[7] But such evidence does not prove that the specific details of the dreams run parallel to these physiological changes: the latter could simply be a non-specific consequence of the entry of consciousness into the dream state.

This point is easier to grasp with the help of an analogy. Consider the interaction between a car and its driver. Under certain conditions, when the car is actually being driven, its movements are closely connected with the actions of the driver, and depend on his perceptions of the road ahead, road signs, dials indicating the internal state of the car, and so on. But under other conditions, this connection is much less close: for example, when the car is stationary with its engine ticking over, the driver might be looking at a map. Although there would be a general relationship between the state of the car and what he was doing – he could not read when driving – there would be no specific connections between the vibrations of the engine and the features of the map he was studying. Likewise the rhythmical electrical activity in the brain need bear no specific relationship to the images experienced in dreams.

If the conscious self has properties of its own which are not reducible to those of matter, energy, morphogenetic fields and motor fields, there is no reason why conscious memories – for example memories of particular past events – need either be stored materially in the brain, *or* depend on morphic resonance. They could well be given directly from past conscious states, across time and space, simply on the basis of similarity with present states. This process would resemble morphic resonance, but differ from it in that it depended not on physical but on conscious states. There would thus be *two* types of long-term memory: motor memory, or habit memory, given by morphic resonance; and conscious memory, given by a direct access of the conscious self to its own past states.[8]

Once the conscious self is admitted to have properties unlike those of any purely physical system, it seems possible that some of these properties might be able to account for parapsychological phenomena which are inexplicable in terms of energetic or of formative causation.[9]

But granted that the self has properties of its own, how does it act upon the body and the external world through the motor fields? There seem to be two ways in which it could do so: first, by selecting between different possible motor fields, causing one course of action to be adopted rather than another; and second, by serving as a creative agency through which new motor fields come into being, for example in 'insight' learning (cf. Section 10.4). In both cases it would act like a formative cause, but one which is, within limits, free and undetermined from the point of view of physical causation. It could indeed be thought of as a formative cause of formative causes.

On this interpretation, consciously controlled actions depend on *three* kinds of causation: conscious causation, formative causation and energetic causation. By contrast, traditional interactionist theories, of the 'ghost in the machine' type, admit only two, conscious and energetic causation, with no formative causation in between. Modified materialism admits a different two, formative and energetic, and denies the existence of conscious causation. And conventional materialism admits only one, energetic causation.[10]

The relationship between conscious causation and formative causation is probably best thought of by analogy with the relationship between formative and energetic causation. Formative causation does not suspend or contradict energetic causation, but imposes a pattern upon events which are indeterminate from an energetic point of view; it selects between energetic possibilities. Likewise, conscious causation does not suspend or contradict formative causation, but selects between motor fields which are equally possible on the basis of morphic resonance.

Situations in which several different patterns of activity are possible might arise either when behaviour under the influence of particular motor fields is not already canalized by innate or habitual chreodes; or when two or more motor fields are competing for control of the body.

In the lower animals, the strong canalization of instinctive patterns of behaviour probably leaves little or no room for conscious causation; but among the higher animals the relatively weak innate canalization of appetitive behaviour may well provide a limited scope. And in man, the enormous range of possible actions gives rise to many ambiguous situations in which conscious choices can be made, both at lower levels, between possible methods of reaching goals already given by the major motor fields, and at higher levels, between competing major motor fields.

On this view, consciousness is primarily directed towards the choice between possible actions, and its evolution has been intimately connected with the increasing scope of conscious causation.

At an early stage in human evolution, this scope must have increased enormously with the development of language, both directly through the capacity to produce an indefinite number of patterns of sounds in the speaking of phrases and sentences; and indirectly through all those actions made possible by this detailed and flexible means of communication. Moreover, in the associated development of conceptual thought, the conscious self must at some stage, in a qualitative leap, have become aware of itself as the agent of conscious causation.

Although conscious creativity reaches its highest development in the human species, it probably also plays an important part in the development of new patterns of behaviour in the higher animals, and may even be of some significance in the lower animals. But conscious causation takes place only within already-established frameworks of formative causation given by morphic resonance from past animals; it cannot account for the major motor fields in the context of which it is expressed, nor can it be regarded as a cause of the characteristic form of the species. Still less can it help to explain the origin of new forms in the plant kingdom. So the problem of evolutionary creativity remains unsolved.

This creativity can either be attributed to a non-physical creative agency which transcends individual organisms; or else it can be ascribed to chance.

The adoption of the latter alternative gives the second of the metaphysical positions compatible with the hypothesis of formative causation, in which the reality of the conscious self as a causal agent is admitted, but the existence of any non-physical agency transcending individual organisms is denied.

12.4 The creative universe

Although a creative agency capable of giving rise to new forms and new patterns of behaviour in the course of evolution would necessarily transcend individual organisms, it need not transcend all nature. It could, for instance, be immanent within life as a whole; in this case it would correspond to what Bergson called the *élan vital*.[11] Or it could be immanent within the planet as a whole, or the solar system, or the entire universe. There could indeed be a hierarchy of immanent creativities at all these levels.

Such creative agencies could give rise to new morphogenetic and motor fields by a kind of causation very similar to the conscious causation considered above. In fact, if such creative agencies are admitted at all, then it is difficult to avoid the conclusion that they must in some sense be conscious selves.

If such a hierarchy of conscious selves exists, then those at higher levels might well express their creativity through those at lower levels. And if such a higher-level creative agency acted through human consciousness, the thoughts and actions to which it gave rise might actually be experienced as coming from an external source. This experience of *inspiration* is in fact well known.

Moreover, if such 'higher selves' are immanent within nature, then it is conceivable that under certain conditions human beings might become directly aware that they were embraced or included within them. And in fact the experience of an inner unity with life, or the earth, or the universe, has often been described, to the extent that it is expressible.

But although an immanent hierarchy of conscious selves might well account for evolutionary creativity within the universe, it could not possibly have given rise to the universe in the first place. Nor could this immanent creativity have any goal if there were nothing beyond the universe towards which it could move. So the whole of nature would be evolving continuously, but blindly and without direction.

This metaphysical position admits the causal efficacy of the conscious self, *and* the existence of creative agencies transcending individual organisms, but immanent within nature. However, it denies the existence of any ultimate creative agency transcending the universe as a whole.

12.5 Transcendent reality

The universe as a whole could have a cause and a purpose only if it were itself created by a conscious agent which transcended it. Unlike the universe, this transcendent consciousness would not be developing towards a goal; it would be its own goal. It would not be striving towards a final form; it would be complete in itself.

If this transcendent conscious being were the source of the universe and of everything within it, all created things would in some sense participate in its nature. The more or less limited 'wholeness' of organisms at all levels of complexity could then be seen as a reflection of the transcendent unity on which they depended, and from which they were ultimately derived.

Thus this fourth metaphysical position affirms the causal efficacy of the conscious self, *and* the existence of a hierarchy of creative agencies immanent within nature, *and* the reality of a transcendent source of the universe.

Notes

1 Some of the modern versions of the philosophy of dialectical materialism would probably provide a good starting point for the development of a modified materialism in this sense. They already include many aspects of the organismic approach, and are based on the idea that reality is inherently evolutionary (Graham, 1972).

2 For a historical account and critical discussion of the various materialist theories, see the chapters by Sir Karl Popper in Popper and Eccles (1977).

3 The hypothesis that both telepathy and memory might be explicable in terms of a new type of trans-temporal and trans-spatial 'resonance' between similar complex systems has in fact already been put forward by Marshall (1960); indeed his suggestion anticipates in several important respects the idea of morphic resonance.

4 Although telepathy and psychokinesis might conceivably be explicable in terms of formative causation, it is difficult to see how this hypothesis could help to account for certain other alleged phenomena, such as clairvoyance, which seem to pose insurmountable problems for any physical theory. For a review of various theories, physical and non-physical, which have been proposed in order to account for the alleged phenomena of parapsychology, see Rao (1977).

5 Ryle (1949).

6 E.g. Eddington (1935); Eccles (1953); Walker (1975).

7 Jouvet (1967).

8 For a discussion of the distinction between motor or habit memory and conscious memory, see Bergson (1911b).

9 See the discussion by Rao (1977).

10 Two different types of dualistic or vitalist theory can be recognized in the light of this classification. The first, exemplified in the writings of Driesch (1908, 1927), postulated the existence of a new type of

causation responsible for repetitive and regular biological processes, corresponding to formative causation in the present sense. The second, developed most brilliantly by Bergson, emphasized conscious causation on the one hand (in his *Matter and Memory*), and evolutionary creativity on the other (in *Creative Evolution*), neither of which could be explained in terms of physical causes.

11 Bergson (1911a).

References

AGAR, W.E., DRUMMOND, F.H. and TIEGS, O.W. (1942) Second report on a test of McDougall's Lamarckian experiment on the training of rats. *Journal of Experimental Biology* 19, 158-167.

AGAR, W.E., DRUMMOND, F.H., TIEGS, O.W. and GUNSON, M.M. (1954) Fourth (final) report on a test of McDougall's Lamarckian experiment on he training of rats. *Journal of Experimental Biology* 31, 307-321.

ANFINSEN, C.B. (1973) Principles that govern the folding of protein chains. *Science* 181, 223-230.

ANFINSEN, C.B. and SCHERAGA, H.A. (1975) Experimental and theoretical aspects of protein folding. *Advances in Protein Chemistry* 29, 205-300.

ASHBY, R.H. (1972) *The Guidebook for the Study of Psychical Research*. Rider, London.

AUDUS, L.J. (1979) Plant geosensors. *Journal of Experimental Botany* 30, 1051-1073.

AYALA, F.J. and DOBZHANSKY, T. (eds) (1974) *Studies in the Philosophy of Biology*. Macmillan, London.

BALDWIN, J.M. (1902) *Development and Evolution.* Macmillan, New York.

BANKS, R.D., BLAKE, C.C.F., EVANS, P.R., HASER, R., RICE, D.W., HARDY, G.W., MERRETT, M. AND PHILLIPS, A.W. (1979) Sequence, structure and activity of phosphoglycerate kinase. *Nature* 279, 773-777.

BELOFF, J. (1962) *The Existence of Mind*. MacGibbon and Kee, London.

BELOFF, J. (1980) Is normal memory a 'paranormal' phenomenon? *Theoria to Theory* 14, 145-161.

BENTLEY, W.A. AND HUMPHREYS, W.J. (1962) *Snow Crystals*. Dover, New York.

BENTRUP, F.W. (1979) Reception and transduction of electrical and mechanical stimuli. In: *Encyclopedia of Plant Physiology* (eds A. Pirson and M.H. Zimmermann), New Series Vol.7, pp. 42-70. Springer-Verlag, Berlin.

BENZER, S. (1973) Genetic dissection of behavior. *Scientific American* 229(6), 24-37.

BERGSON, H. (1911a) *Creative Evolution*. Macmillan, London.

BERGSON, H. (1911b) *Matter and Memory*. Allen and Unwin, London.

BOHM, D. (1969) Some remarks on the notion of order. In: Waddington (ed.) (1969).

BOHM, D. (1980) *Wholeness and the Implicate Order*. Routledge and Kegan Paul, London.

BONNER, J.T. (1958) *The Evolution of Development*. Cambridge University Press, Cambridge.

BOSE, J.C. (1926) *The Nervous Mechanism of Plants*. Longmans, Green & Co., London.

BOYCOTT, B.B. (1965) Learning in the octopus. *Scientific American* 212(3), 42-50.

BRENNER, S. (1973) The genetics of behaviour. *British Medical Bulletin* 29, 269-271.

BROADBENT, D.E. (1961) *Behaviour*. Eyre and Spottiswoode, London.

BROWN, J.L. (1975) *The Evolution of Behavior*. Norton, New York.

BÜNNING, E. (1973) *The Physiological Clock*. English Universities Press, London.

BURGESS, J. AND NORTHCOTE, D.H. (1968) The relationship between the endoplasmic reticulum and microtubular aggregation and disaggregation. *Planta* 80, 1-14.

BURR, H.S. (1972) *Blueprint for Immortality*. Neville Spearman, London.

BURSEN, H.A. (1978) *Dismantling the Memory Machine*. Reidel, Dordrecht.

BUTLER, S. (1878) *Life and Habit*. Cape, London.

CARINGTON, W. (1945) *Telepathy*. Methuen, London.

CLARKE, R. (1980) Two men and their dogs. *New Scientist* 87, 303-304.

CLOWES, F.A.L. (1961) *Apical Meristems*. Blackwell, Oxford.

CREW, F.A.E. (1936) A repetition of McDougall's Lamarckian experiment. *Journal of Genetics* 33, 61-101.

CRICK, F.H.C. (1967) *Of Molecules and Men*. University of Washington Press, Seattle.

CRICK, F.H.C. AND LAWRENCE, P. (1975). Compartments and polyclones in insect development. *Science* 189, 340-347.

CRICK, F.H.C. AND ORGEL, L. (1973) Directed panspermia. *Icarus* 10, 341-346.

CURRY, G.M. (1968) Phototropism. In: *Physiology of Plant Growth and Development* (ed. M.B. Wilkins). McGraw-Hill, London.

DARWIN, C. (1875) *The Variation of Animals and Plants Under Domestication*. Murray, London.

DARWIN, C. (1880) *The Power of Movement in Plants*. Murray, London.

Darwin, C. (1882) *The Movements and Habits of Climbing Plants.* Murray, London.

Dawkins, R. (1976) *The Selfish Gene.* Oxford University Press, Oxford.

De Chardin, P.T. (1959) *The Phenomenon of Man.* Collins, London.

D'Espagnat, B. (1976) *The Conceptual Foundations of Quantum Mechanics.* Benjamin, Reading, Mass.

Dilger, W.C. (1962) The behavior of lovebirds. *Scientific American* **206**, 88-98.

Dostal, R. (1967) *On Integration in Plants.* Harvard University Press, Cambridge, Mass.

Driesch, H. (1908, second edition 1929) *Science and Philosophy of the Organism.* A. & C. Black, London.

Driesch, H. (1914) *History and Theory of Vitalism.* Macmillan, London.

Driesch, H. (1927) *Mind and Body.* Methuen, London.

Duncan, R. and Weston-Smith, M. (eds.) (1977) *Encyclopedia of Ignorance.* Pergamon Press, Oxford.

Dustin, P. (1978) *Microtubules.* Springer-Verlag, Berlin.

Eccles, J.C. (1953) *The Neurophysiological Basis of Mind.* Oxford University Press, Oxford.

Eckert, R. (1972) Bioelectric control of ciliary activity. *Science* **176**, 473-481.

Eddington, A. (1935) *The Nature of the Physical World.* Dent, London.

Eigen, M. and Schuster, P. (1979) *The Hypercycle.* Springer-Verlag, Heidelberg and New York.

Elsasser, W.M. (1958) *Physical Foundations of Biology.* Pergamon Press, London.

Elsasser, W.M. (1966) *Atom and Organism.* Princeton University Press, Princeton.

Elsasser, W.M. (1975) *The Chief Abstractions of Biology.* North Holland, Amsterdam.

Emmet, D. (1966) *Whitehead's Philosophy of Organism.* Macmillan, London.

Fisher, R.A. (1930) *Genetical Theory of Natural Selection.* Clarendon Press, London.

Goebel, K. (1898) *Organographie der Pflanzen.* Fischer, Jena.

Goldschmidt, R. (1940) *The Material Basis of Evolution.* Yale University Press, New Haven.

Goodwin, B.C. (1979) On morphogenetic fields. *Theoria to Theory* 13, 109-114.

Gould, S.J. (1980) Return of the hopeful monster. In: *The Panda's Thumb.* Norton, New York.

GRAHAM, L.A. (1972) *Science and Philosophy in the Soviet Union.* Knopf, New York.

GURWITSCH, A. (1922) Über den Begriff des embryonalen Feldes. *Archiv für Entwicklungsmechanik* 51, 383-415.

HAKEN, H. (1977) *Synergetics.* Springer-Verlag, Berlin.

HALDANE, J.B.S. (1939) The theory of the evolution of dominance. *Journal of Genetics* 37, 365-374.

HARAWAY, D.J. (1976) *Crystals, Fabrics and Fields.* Yale University Press, New Haven.

HARDY, A. (1965) *The Living Stream.* Collins, London.

HASTED, J.B. (1978) Speculations about the relation between psychic phenomena and physics. *Psychoenergetic Systems* 3, 243-257.

HEBB, D.O. (1949) *The Organization of Behavior.* Wiley, New York.

HESSE, M.B. (1961) *Forces and Fields.* Nelson, London.

HILEY, B.J. (1980) Towards an algebraic description of reality. *Annales de la Fondation Louis de Broglie* 5, 75-103.

HINDE, R.A. (1966) *Animal Behavior.* McGraw-Hill, New York.

HINGSTON, R.W.G. (1928) *Problems of Instinct and Intelligence.* Arnold, London.

HOLDEN, A. AND SINGER, P. (1961) *Crystals and Crystal Growing.* Heinemann, London.

HOYLE, F. AND WICKRAMASINGHE, C. (1978) *Lifecloud.* Dent, London.

HUXLEY, J. (1942) *Evolution: The Modern Synthesis.* Allen and Unwin, London.

HUXLEY, T.H. (1867) *Hardwicke's Science Gossip* 3, 74.

JAFFE, M.J. (1973) Thigmomorphogenesis. *Planta* 114, 143-157.

JENNINGS, H.S. (1906) *Behavior of the Lower Organisms.* Columbia University Press, New York.

JENNY, H. (1967) *Cymatics.* Basileus Press, Basel.

JOUVET, M. (1967) The states of sleep. *Scientific American* 216(2), 62-72.

JUNG, C.G. (1959) *The Archetypes and the Collective Unconscious.* Routledge and Kegan Paul, London.

KAMMERER, P. (1924) *The Inheritance of Acquired Characteristics.* Boni and Liveright, New York.

KANDEL, E.R. (1979) Small systems of neurons. *Scientific American* 241(3), 61-71.

KATZ, B. (1966) *Nerve, Muscle and Synapse.* McGraw-Hill, New York.

KATZ, B. AND MILEDI, R. (1970) Membrane noise produced by acetylcholine. *Nature* 226, 962-963.

KING, M.C. AND WILSON, A.C. (1975) Evolution at two levels in humans and chimpanzees. *Science* **188**, 107-116.

KOESTLER, A. (1967) *The Ghost in the Machine*. Hutchinson, London.

KOESTLER, A. (1971) *The Case of the Midwife Toad*. Hutchinson, London.

KOESTLER, A. AND SMYTHIES, J.R. (eds) (1969) *Beyond Reductionism*. Hutchinson, London.

KÖHLER, W. (1925) *The Mentality of Apes*. Harcourt Brace, New York.

KRSTIC, R.V. (1979) *Ultrastructure of the Mammalian Cell*. Springer-Verlag, Berlin.

KUHN, T.S. (1962) *The Structure of Scientific Revolutions*. Chicago University Press, Chicago.

LASHLEY, K.S. (1950) In search of the engram. *Symposia of the Society for Experimental Biology* **4**, 454-482.

LAWDEN, D.F. (1980) Possible psychokinetic interactions in quantum theory. *Journal of the Society for Psychical Research* **50**, 399-407.

LECLERC, I. (1972) *The Nature of Physical Existence*. Allen and Unwin, London.

LENARTOWICZ, P. (1975) *Phenotype-Genotype Dichotomy*. Gregorian University, Rome.

LEWIS, E.B. (1963) Genes and developmental pathways. *American Zoologist* **3**, 33-56.

LEWIS, E.B. (1978) A gene complex controlling segmentation in *Drosophila*. *Nature* **276**, 565-570.

LINDAUER, M. (1961) *Communication Among Social Bees*. Harvard University Press, Cambridge, Mass.

LOIZOS, C. (1967) Play behaviour in higher primates: a review. In: *Primate Ethology* (ed. D. Morris). Weidenfeld and Nicolson, London.

MACKIE, J.L. (1974) *The Cement of the Universe*. Oxford University Press, Oxford.

MACKINNON, D.C. AND HAWES, R.S.J. (1961) *An Introduction to the Study of Protozoa*. Oxford University Press, Oxford.

MACWILLIAMS, H.K. AND BONNER, J.T. (1979) The prestalk-prespore pattern in cellular slime moulds. *Differentiation* **14**, 1-22.

MAHESHWARI, P. (1950) *An Introduction to the Embryology of Angiosperms*. McGraw-Hill, New York.

MANNING, A. (1975) Behaviour genetics and the study of behavioural evolution. In: *Function and Evolution in Behaviour* (eds G.P. Baerends, C. Beer and A. Manning). Oxford University Press, Oxford.

MARAIS, E. (1971) *The Soul of the White Ant*. Cape and Blond, London.

MARSHALL, N. (1960) ESP and memory: a physical theory. *British Journal for the Philosophy of Science* **10**, 265-286.

MASTERS, M.T. (1869) *Vegetable Teratology*. Ray Society, London.

MAYR, E. (1963) *Animal Species and Evolution*. Harvard University Press, Cambridge, Mass.

McDOUGALL, W. (1927) An experiment for the testing of the hypothesis of Lamarck. *British Journal of Psychology* **17**, 267-304.

McDOUGALL, W. (1930) Second report on a Lamarckian experiment. *British Journal of Psychology* **20**, 201-218.

McDOUGALL, W. (1938) Fourth report on a Lamarckian experiment. *British Journal of Psychology* **28**, 321-345.

MEDAWAR, P.B. (1968) *The Art of the Soluble*. Methuen, London.

MEDVEDEV, Z.A. (1969) *The Rise and Fall of T.D. Lysenko*. Columbia University Press, New York.

MEINHARDT, H. (1978) Space-dependent cell determination under the control of a morphogen gradient. *Journal of Theoretical Biology* **74**, 307-321.

MONOD, J. (1972) *Chance and Necessity*. Collins, London.

MORATA, G. AND LAWRENCE, P.A. (1977) Homoeotic genes, compartments and cell determination in *Drosophila*. *Nature* **265**, 211-216.

NEEDHAM, J. (1942) *Biochemistry and Morphogenesis*. Cambridge University Press, Cambridge.

NEMETHY, G. AND SCHERAGA, H.A. (1977) Protein folding. *Quarterly Review of Biophysics* **10**, 239-352.

NICOLIS, G. AND PRIGOGINE, I. (1977) *Self-Organization in Nonequilibrium Systems*. Wiley-Interscience, New York.

PARSONS, P.A. (1967) *The Genetic Analysis of Behaviour*. Methuen, London.

PAULING, L. (1960) *The Nature of the Chemical Bond* (third edition). Cornell University Press, Ithaca.

PEARSON, K. (1924) *Life of Francis Galton*. Cambridge University Press, Cambridge.

PECHER, C. (1939) La fluctuation d'excitabilité de la fibre nerveuse. *Archives Internationales de Physiologie* **49**, 129-152.

PENZIG, O. (1921-2) *Pflanzen-Teratologie*. Borntraeger, Berlin.

PICKETT-HEAPS, J.D. (1969) The evolution of the mitotic apparatus. *Cytobios* **3**, 257-280.

PICKETT-HEATS, J.D. (1975) *Green Algae*. Sinauer Associates, Sunderland, Mass.

POLANYI, M. (1958) *Personal Knowledge*. Routledge and Kegan Paul, London.

POPPER, K.R. (1965) *Conjectures and Refutations*. Routledge and Kegan Paul, London.

POPPER, K.R. (1967) Quantum mechanics without 'the observer'. In: M.

Bunge (ed.) *Quantum Theory and Reality*. Springer Verlag, Berlin.
POPPER, K.R. AND ECCLES, J.C. (1977) *The Self and its Brain*. Springer International, Berlin.
PRIBRAM, K.H. (1971) *Languages of the Brain*. Prentice Hall, Englewood Cliffs.

RAO, K.R. (1977) On the nature of psi. *Journal of Parapsychology* 41, 294-351.
RAPP, P.E. (1979) An atlas of cellular oscillations. *Journal of Experimental Biology* 81, 281-306.
RAVEN, P.H., EVERT, R.F. AND CURTIS, H. (1976) *Biology of Plants*. Worth Publishers, Inc., New York
RENSCH, B. (1959) *Evolution Above the Species Level*. Methuen, London.
RHINE, J.B. AND MCDOUGALL, W. (1933) Third report on a Lamarckian experiment. *British Journal of Psychology* 24, 213-235.
RICARD, M. (1969) *The Mystery of Animal Migration*. Constable, London.
RIEDL, R. (1978) *Order in Living Organisms*. Wiley Interscience, Chichester and New York.
RIGNANO, E. (1926) *Biological Memory*. Harcourt, Brace and Co., New York.
ROBERTS, K. AND HYAMS, J.S. (eds) (1979) *Microtubules*. Academic Press, London.
ROBLIN, G. (1979) *Mimosa pudica*: a model for the study of the excitability in plants. *Biological Reviews* 54, 135-153.
RUSSELL, B. (1921) *Analysis of Mind*. Allen and Unwin, London.
RUSSELL, E.S. (1945) *The Directiveness of Organic Activities*. Cambridge University Press, Cambridge.
RUYER, R. (1974) *La Gnose de Princeton*. Fayard, Paris.
RYLE, G. (1949) *The Concept of Mind*. Hutchinson, London.

SATTER, R.L. (1979) Leaf movements tendril curling. In: *Encyclopedia of Plant Physiology* (eds A. Pirson and M.H. Zimmermann), New Series Vol. 7, pp. 442-484. Springer-Verlag, Berlin.
SCHOPENHAUER, A. (1883) *The World as Will and Idea*, Book 1, Section 7. Kegan Paul, London.
SEMON, R. (1912) *Das Problem der Vererbung Erworbener Eigenschaften*. Engelmann, Leipzig.
SEMON, R. (1921) *The Mneme*. Allen and Unwin, London.
SERRA, J.A. (1966) *Modern Genetics* Vol. II, pp.269-270. Academic Press, London.
SHELDRAKE, A.R. (1973) The production of hormones in higher plants. *Biological Reviews* 48, 509-559.
SHELDRAKE, A.R. (1974) The ageing, growth and death of cells. *Nature* 250, 381-385.
SHELDRAKE, A.R. (1980a) Three approaches to biology. I. The mechanistic

theory of life. *Theoria to Theory* 14, 125-144.

SHELDRAKE, A.R. (1980b) Three approaches to biology. II. Vitalism. *Theoria to Theory* 14, 227-240.

SHELDRAKE, A.R. (1981) Three approaches to biology. III. Organicism. *Theoria to Theory* 14, 301-311.

SIEGELMAN, H.W. (1968) Phytochrome. In: *Physiology of Plant Growth and Development* (ed. M.B. Wilkins). McGraw-Hill, London.

SINNOTT, E.W. (1963) *The Problem of Organic Form*. Yale University Press, New Haven.

SKINNER, B.F. (1938) *The Behavior of Organisms*. Appleton Century, New York.

SLEIGH, M.A. (1968) Co-ordination of the rhythm of beat in some ciliary systems. *International Review of Cytology* 25, 31-54.

SNOAD, B. (1974) A preliminary assessment of 'leafless peas'. *Euphytica* 23, 257-265.

SPEAR, N.E. (1978) *The Processing of Memories*. Lawrence Erlbaum Associates, Hillsdale, N.J.

STEBBINS, G.L. (1974) *Flowering Plants: Evolution Above the Species Level*. Harvard University Press, Cambridge, Mass.

STEVENS, C.F. (1977) Study of membrane permeability changes by fluctuation analysis. *Nature* 270, 391-396.

STREET, H.E. AND HENSHAW, G.G. (1965) Introduction and methods employed in plant tissue culture. In: *Cells and Tissues in Culture* (ed. E.N. Willmer) Vol. 3, pp.459-532. Academic Press, London.

SUPPES, P. (1970) *A Probabilistic Theory of Causality*. North Holland, Amsterdam.

TAYLOR, J.G. AND BALANOVSKI, E.(1979) Is there any scientific explanation of the paranormal? *Nature* 279, 631-633.

THOM. R. (1975a) *Structural Stability and Morphogenesis*. Benjamin, Reading, Mass.

THOM, R. (1975b) D'un modèle de la science a une science des modèles. *Synthèse* 31, 359-374.

THOMPSON, D'ARCY W. (1942) *On Growth and Form*. Cambridge University Press, Cambridge.

THORPE, W.H. (1963) *Learning and Instinct in Animals* (second edition). Methuen, London.

THORPE, W.H. (1978) *Purpose in a World of Chance*. Oxford University Press, Oxford.

THOULESS, R.H. (1972) *From Anecdote to Experiment in Psychical Research*. Routledge and Kegan Paul, London.

TINBERGEN, N. (1951) *The Study of Instinct*. Oxford University Press, Oxford.

VERVEEN, A.A. AND DE FELICE, L.J. (1974) Membrane noise. *Progress in Biophysics and Molecular Biology* 28, 189-265.
VON BERTALANFFY, L. (1933) *Modern Theories of Development*. Oxford University Press, London.
VON BERTALANFFY, L. (1971) *General Systems Theory*. Allen Lane, London.
VON FRISCH, K. (1975) *Animal Architecture*. Hutchinson, London.

WADDINGTON, C.H. (1957) *The Strategy of the Genes*. Allen and Unwin, London.
WADDINGTON, C.H. (ed.) (1969) *Towards a Theoretical Biology. 2: Sketches*. Edinburgh University Press, Edinburgh.
WADDINGTON, C.H. (1975) *The Evolution of an Evolutionist*. Edinburgh University Press, Edinburgh.
WALKER, E.H. (1975) Foundations of paraphysical and parapsychological phenomena. In: *Quantum Physics and Parapsychology* (ed. L.Otera). Parapsychology Foundation, New York.
WARDLAW, C.W. (1965) *Organization and Evolution in Plants*. Longmans, London.
WATSON, J.B. (1924) *Behaviorism*. Chicago University Press, Chicago.
WEISS, P. (1939) *Principles of Development*. Holt, New York.
WHITEHEAD, A.N. (1928) *Science and the Modern World*. Cambridge University Press, Cambridge.
WHITEMAN, J.H.M. (1977) Parapsychology and physics. In: Wolman (ed.) (1977).
WHYTE, L.L. (1949) *The Unitary Principle in Physics and Biology*. Cresset Press, London.
WIGGLESWORTH, V.B. (1964) *The Life of Insects*. Weidenfeld and Nicolson, London.
WIGNER, E. (1961) Remarks on the mind-body question. In: *The Scientist Speculates* (ed. I.J. Good). Heinemann, London.
WIGNER, E. (1969) Epistemology in quantum mechanics. In: *Contemporary Physics: Trieste Symposium* 1968. Vol. II, pp. 431-438. International Atomic Energy Authority, Vienna.
WILLIAMS, R.J.P. (1979) The conformational properties of proteins in solution. *Biological Reviews* 54, 389-437.
WILLIS, J.C. (1940) *The Course of Evolution*. Cambridge University Press, Cambridge.
WILLMER, E.N. (1970) *Cytology and Evolution* (second edition). Academic Press, London.
WILSON, E.O. (1975) *Sociobiology: The New Synthesis*. Harvard University Press, Cambridge, Mass.
WOLFF, G. (1902) *Mechanismus und Vitalismus*. Leipzig.

WOLMAN, B.B. (ed.) (1977) *Handbook of Parapsychology*. Van Nostrand Reinhold, New York.

WOLPERT, L. (1978) Pattern formation in biological development. *Scientific American* 239 (4), 154-164.

WOODGER, J.H. (1929) *Biological Principles*. Kegan Paul, Trench, Trubner & Co., London.

Index of Names

Index of Subjects